Animal Homing and Migration through Biogeomagnetic Beaming

A Revealing Study into Animal Orientation Phenomena as Induced Lateral Directional Equilibrium Reflex

Major J. Huber

Sci-Tech Communication Services
Encinitas, California 92024

Published 1993. Printed in the United States of America

Edited by Becky Colgan

Library of Congress Catalog Card Number
93-083838

ISBN No. 0-9635392-0-5

To my family,

whose encouragement and assistance

are reflected in every page.

Table of Contents

Preface v

Author's Note vii

1 Biogeomagnetic Beaming 1

2 Biogeomagnetic Beaming Procedure through Bioreflex Action 7

Basic beaming procedure 7
Proper beaming procedure 12
Determining the beam direction and intensity ... 15
BRI variations 17

3 Directional Beams 21

The migration beam (northern hemisphere) 22
The polar light dominance factor 25
The homing beam 28
Color-value factor 37
Color value in relation to bird homing and migration 40
Shoreline beam 42
Beamer's directional antenna influence factor ... 51

4 Directional Beaming Indicator 57

Indicator design 57
Beamer reflex guidance factor and its relation to animal homing and migration 65
Human indicator beam detection versus natural animal directional beaming 67
Relation to bird homing and migration (theoretical) 70
Bird on a beam 74

5 Interaction of Biogeomagnetic Forces 87

The LMR apparatus (light, magnet, rotation principles applied) 87
LMR design and operation 90
Light/magnetic field interactions 91
The georotational polar blink factor 93
Application of rotation factor to biobeaming 100
Pivot-to-ground factor 103
Local disruptive biobeaming influences 104

6 Bineural Factor 109

Bineural factor defined 109
Bineural polarity factor 112
Conclusion 114

Appendix 115

Author's Note 115
Miscellaneous related data of research value 115

Bibliography 119

Index 120

List of Figures

1a Human adaption for directional beam detection (facing view) showing on-course 0% BRI reflex action 10

1b Human adaption for directional beam detection (facing view) showing off-course 100% BRI reflex action 11

2 Seasonal migration beam 23

3 Bar magnet 26

4 Homing beam 30

5 Biobeaming 32

6 Homing beam source emissions 34

7 Overlapping homing beam sources/emissions 36

8 Cross-section of shoreline beam 43

9 Shoreline beam in relation to migration beam orientation 46

10 Water beams versus land beams (shoreline area) .. 48

11 Beaming indicators 59

12 XA 175° vee indicator versus 87.5° vee indicator .. 61

13 Vee angle design 64

14 LMR apparatus (cutaway view) 89

15 LMR test data 98

Preface

Throughout history, dedicated scientists have spent countless hours pursuing a better understanding of animal homing and migration phenomena. In recent years the discovery that magnetic fields can affect the inherent homing abilities of homing pigeons was a significant step forward in this field of science.

In 1964 I inadvertently stumbled upon a similar discovery while investigating a nonrelated problem. This discovery launched me on a quest that, so far, represents more than twenty-five years of exhaustive investigation and study. This book presents the research and findings that evolved to reveal the existence of the unique and complex phenomenon I refer to as *biogeomagnetic lateral directional beaming* (or *biobeaming*).

The ability to detect, through one's own reflexes, the existence of directional beaming forces is not a unique experience. Any investigative individual can quickly learn to detect these forces through effortless basic procedures. I began to understand the subtle characteristics of directional beaming phenomena after an extended period of repeated beaming procedure testing. I can now confidently state that detection of such beams can be a normal function in the human animal. However, I believe the detection sensitivity has been dulled through human evolution. I also believe a finely honed lateral directional beaming ability has been repeatedly bred into many other animals, since only those so gifted could survive to perpetuate the species.

Although the research effort put into this work has been extensive, it does little more than scratch the surface in understanding biogeomagnetic phenomena. The objective of this book is to reveal the existence of beaming forces and to illustrate how their characteristics are closely reflected in the known patterns of animal homing and migration behavior.

I do not intend to decry present theories that suggest the migration bird is capable of executing terrestrial navigational feats that humans cannot duplicate without sophisticated equipment and training. My findings are, however, a revolutionary approach to the study of animal migration and homing orientation that substantiates the recent suspicion that pigeons, for example, do indeed *beam* toward their home loft.

Past investigators of animal homing and migration phenomena entered this dark, complex tunnel by studying the gifted wild animals. Through an accidental discovery, I entered this same tunnel from the other end and found myself studying animals' directional orientation using *human* animals instead.

I hope the biogeomagnetic beaming data revealed in this book will stir the curiosity and reasoning of those truly dedicated to better understanding the mysteries of animal homing and migration. It will surely open a Pandora's box of new questions and speculation in many fields of science as well as biogeomagnetic directional beaming.

Author's Note

Preparation of this report has been a challenge. Writing is neither my forté nor a great love of mine. Nevertheless, this book has evolved out of sheer urgency to print the data revealed in my studies of animal orientation phenomena. This effort is not intended to bring materialistic reward. The true purpose of this book stems from a sense of incurable optimism that prompts me to speculate that someone, somewhere, will recognize the potential of this new field of research.

Letters, comments, and questions are welcome. Write to

Major J. Huber
P. O. Box 1742
Twenty-Nine Palms, CA 92277

1 Biogeomagnetic Beaming

My twenty-plus years of relentless investigations have evolved from an inadvertent discovery that led to a well-reinforced theory that the homing and migration bird, for example, is laterally guided by a beam of geomagnetic influence. This subtle and involuntary action affects the animal's neuroreflexes, possibly inducing a subtle sense of directional security when facing or traveling in one specific lateral direction and ceasing to exist when veering from such a course. This reaction is comparable to what humans experience, even as infants when not in a state of vertical equilibrium. A warning sensation of discomforting vertigo occurs. The bird on a migration or homing course would, in addition, sense a parallel reaction referred to in this report as *lateral directional equilibrium.*

This book presents the basic findings that have materialized from my investigations. It is hoped that some dedicated individuals may recognize the potential of this unique field of study and become interested in probing more deeply into the phenomena of animal homing migration.

All statements pertaining to biobeaming phenomenon are based on recorded results of actual beaming tests repeated over the years. Any correlation of beaming findings with that of bird homing and migration feats are theoretical, although such correlation is hard to escape, considering the close parallel between the discovered biogeomagnetic beaming patterns and known bird homing and migration patterns. Applied syllogistics can hardly be avoided.

This study deals with existing forces, the interaction of which trigger other forces. The following basic factors have been found to be involved:

- Earth's magnetic field (geomagnetics)
- Electromagnetic forces (light)
- Biological reflex action (neural)
- Rotation factor

The phenomenon resulting from the interaction of these forces is referred to as *lateral directional beams*. Its detection is referred to as *biogeomagnetic beaming*.

When certain geomagnetic forces interact with light and certain other biological forces under given circumstances, the resulting subtle reflex action of the individual being tested can be visually and physically detected. This reflex (or *beaming* reaction) varies in intensity depending on which lateral and frontal direction the test individual (*beamer*) assumes. It is a well-established fact that a consistent pattern exists relative to such beamer orientations, inducing varied degrees of a particular reflex action intensity ranging from 100% beamer's reflex intensity (BRI) to 0% BRI. This reaction is a physical result and is therefore unquestionably recognized.

In the course of early beaming test investigations, my initial discovery (1964) showed that a strong 100% BRI reaction occurred when facing southeast. The 0% BRI reaction was detected when facing northwest.

Although investigation had not *yet* established a relationship between the beaming reactions and animal homing and migration phenomena, this directional beam[1] so strongly suggested a guiding force that the term *migration beam* was adopted for identification.

Continued northern hemispheric investigations of this unique directional beam reaction led, eventually, to disappointment. Later tests indicated that the northwest beam, originally discovered, ceased to exist.[2]

However, a positive (unnatural) 50% BRI reflex action persisted in all directions laterally faced. Such abnormality in reflex action was a puzzling phenomenon prompting several years of subsequent investigation. The findings that emerged revealed the existence of a consistent pattern of seasonal, northern hemispheric, directional beam variations.[3]

When proper procedure was observed (in northern hemispheric tests), the following phenomena occurred:

- Following the September equinox, a direction of 0% BRI reflex can be detected only when facing southeast.

---

[1] The word *beam* in this report refers to the existing biogeomagnetic influence that acts upon a beamer's neuroreflexes.

[2] The period between the summer solstice and autumnal equinox, when no migration beams can be detected, was eventually found to be responsible for the negative beam reaction loss.

[3] All beaming tests in this summary were conducted in the northern hemisphere. No research has been done, to date, in the southern hemisphere.

- Following the March equinox, a direction of 0% BRI reflex can be detected only when facing northwest.
- Between the solstices and equinoxes, no specific direction of 0% BRI reflex is detected.

In subsequent years of investigating the phenomena of seasonal directional beaming, additional discoveries were made. The results of this study terminated with the realization that not one directional beam existed, but three such beams with individual characteristics (described in Chapter 3). The discovery of a directional beam influence demonstrated a relationship to many unexplained migration and homing feats of the animal world. For this reason, the following individual directional beams were identified:

- Migration beam
- Homing beam
- Shoreline (or water) beam

This book describes each beam in detail to illustrate their characteristic differentials, and yet their relation to one another.

An equally important topic that is discussed is the most critical development in the overall biobeaming project. This development involves the use of a specifically designed testing apparatus to produce simulated seasonal directional beam forces detectable through the bioreflexes. This test instrument is the Light-Magnet-Rotation (LMR) apparatus, which employs the same geomagnetic, electromagnetic factors found to trigger Earth's normal seasonal migration beams as well as the necessary role played by the *rotation factor*.

This book describes the following in detail:

- The *biogeomagnetic lateral directional beaming* procedure required to detect these directional forces
- The individual characteristics of the migration beam, homing beam, and shoreline beam
- Bioreflex action, which is a biological, physical reaction to directional beam forces referred to as *biogeomagnetic beaming*

In addition, the theories on interacting forces involved and responsible for these beaming forces have been applied in the construction of the LMR test apparatus. The LMR was indispensable in further investigating the beaming phenomena because of its ability to duplicate at any season the seasonal geomagnetic migration beam forces, thereby expediting their study. Of no less importance, LMR, as a working apparatus, substantiates the theories behind its design. LMR apparatus capabilities and design are explained in Chapter 5.

The final chapters of this book detail the bioreflex phenomenon during beaming procedure along with other findings related to the *bineural factor*.

Throughout this book you will find limited analogies that apply to the subject of animal homing and migration through beaming to demonstrate the existing parallel between biogeomagnetic beaming patterns and known animal homing and migration patterns.

Persistent research of biobeaming phenomena resulted in an accumulation of thousands of pages of notes upon which this report is based. This limited text cannot possibly reflect all findings recorded; however, it does detail all major facets found to exist in the complex study of a beamer's lateral directional orientation through biogeomagnetic beaming procedures.

2 Biogeomagnetic Beaming Procedure through Bioreflex Action

The information presented in this chapter is expanded upon in subsequent chapters. This overview introduces important concepts that are referred to throughout the book.

Basic beaming procedure

Introducing directional biogeomagnetic beaming phenomena by first explaining the basic beaming procedure is especially important. Human interest in pursuing this subject rests heavily upon the experience of realizing that natural forces do exist that subtly act upon certain bioreflexes. With credibility established, acceptance and interest follow.

The findings in this book are the product of countless beaming test results that can be detected no other way than through bioreflex action, triggered by a specific uncomplicated beaming procedure.

The directional beaming procedure occurs when an individual detects a given existing directional beam reaction. The procedure, conducted by a *beamer,* demands no equipment. However, to increase the intensity of beam detection, a simple instrument is used, thereby rendering a more *perceptual* reflex action. This *beaming indicator* serves three basic purposes:

1. It is a link between a beamer's left and right bineural systems. (See Chapter 6.)
2. It is a directional antenna for detecting the presence and orientation of existing directional beams.
3. It is an indicator for calibrating the percent of beamer's reflex intensity (BRI).

Beaming indicator design and proper usage must follow specific requirements to be a factor in inducing reflex action in a beamer, interpreted as *directional beam detection.* (See Chapter 4.)

The beaming indicator, a plain, curved instrument (discussed in Chapter 4), is held by the beamer with natural, slightly inward gravity-induced pressure of one finger of each hand. It is held with its arc downward. With elbows held outward (akimbo) and arms totally relaxed at shoulder joints, the indicator should hang centered with the vertical line of the body at *approximately waist height.* Figures 1a and 1b illustrate the proper hold on a beaming indicator during the biobeaming procedure. Note the bineural effect. Properly held, the indicator is free to swing like a pendulum, enabling a repeated, slight "cocking" action either to the beamer's left or right. This repeated cocking action affects the presence of any beamer's reflex action, in that a locking or toggling reflex is triggered when a positive beam force is encountered. This involuntary locking through reflex action is sensed to act upon the scapular area of the human beamer.[1] If the beaming indicator, properly held, is freely swung to the right (for example) while being gently, repeatedly cocked in that direction, it will tend not to return to the natural center position of a beamer's body. Instead, it will lock to the right in an off-centered (positive) position by as much as eight inches (using a 22-inch indicator span). (See Figure 1b.)

A positive (100% BRI) reaction induces the maximum unnatural toggling action (i.e., off-course).[2] It is not, however, the maximum BRI that represents the guiding beam direction, but that of the lesser 0% BRI (on-course) no-reflex action. This reaction is triggered when the indicator's broadside is oriented to face in a minimum reaction on-course direction (180 degrees from the maximum off-course direction). At this point, the degree of the

---

[1] Such a "locking-in" reaction could well be reasoned to somewhat parallel the recognized reflex action that involuntarily "locks" the knees of animals in a standing position. This bioreflex "locking-in" is immediately overcome, however, by the slightest voluntary suggestion to do so.

[2] The positive (locking-in) reflex action is quite evident through calibrations on the beaming indicator. However, many past errors have repeatedly proven that this reflex action is of a most sensitive nature. Careless, hurried beaming tests can critically distort accuracy of results. For example, should a beamer, in testing, expect a given lateral direction to induce a lesser degree of reflex action (on-course)? Only careful, proper beaming procedure can prevent this, which is best accomplished by gently recocking the beaming indicator to a slight off-center position and totally relaxing all shoulder joint muscles.

Inversely, the possibility of subconsciously *holding* the indicator in an off-center position is far less subtle. In this case, a beamer is aware of any self-imposed force required to combat gravity's influence to return the weight of the beamer's arms and the indicator to a natural position.

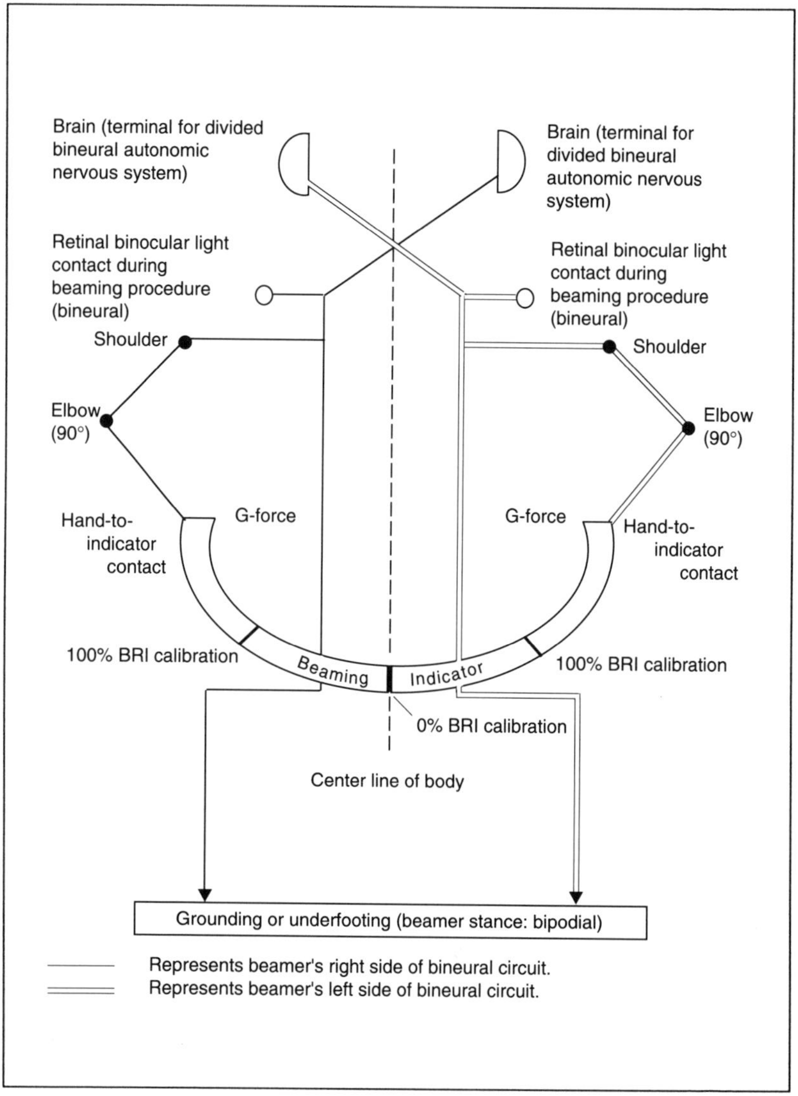

Figure 1a. **Human adaption for directional beam detection (facing view) showing on-course 0% BRI reflex action**

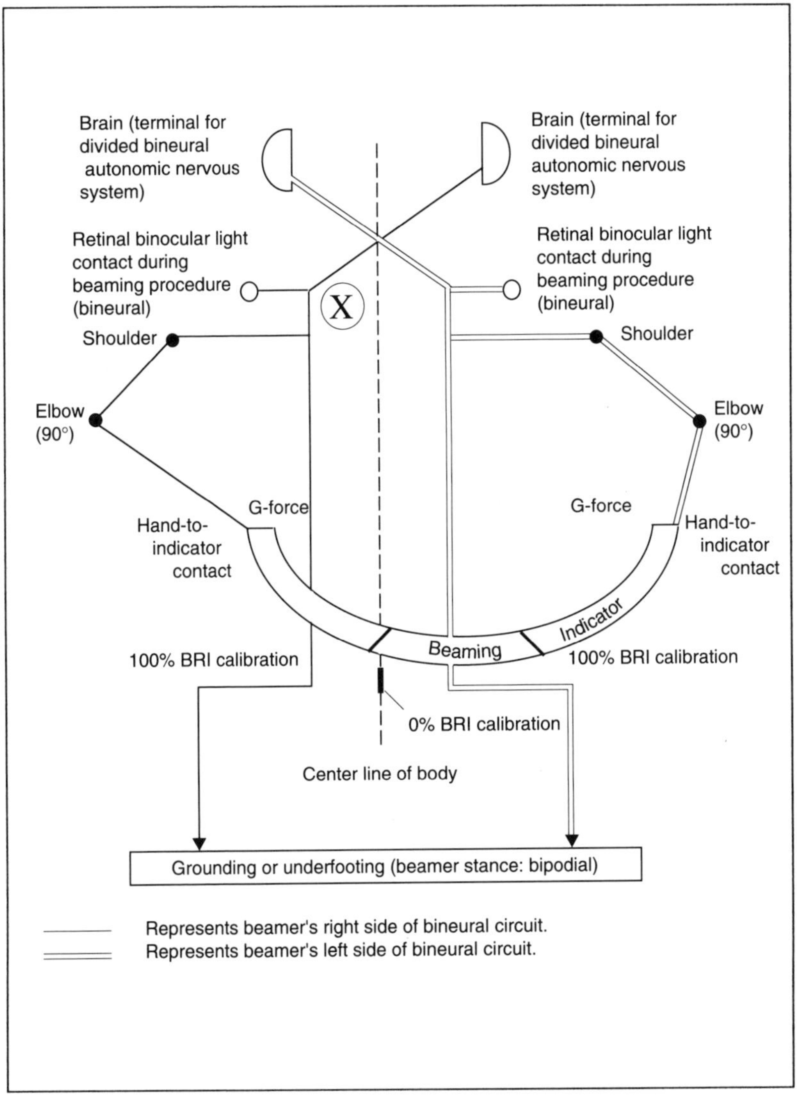

Figure 1b. Human adaption for directional beam detection (facing view) showing off-course 100% BRI reflex action

Proper beaming procedure

Hold indicator at waist height, shoulder joints relaxed, elbows at 90-degree angle, allowing loose pendular action to arms. Hands should hold indicator with inward force (arrows). Let the beaming indicator fall to a relaxed, natural position with the center of the indicator aligned with the center line of the body (indicating an on-course, 0% BRI reflex action). The result is an inward pressure (g-force) at the indicator's outward tips due to proper beamer procedure, creating stress on the indicator's curved design.

Figure 1a shows the indicator in a relaxed, natural position with the center of the indicator aligned with the center line of the body.

Figure 1b shows the beaming indicator in an off-center position. This is due to the reflex and locking action of a beamer induced when oriented in a lateral direction other than toward an on-course orientation. The X represents the approximate area in the beamer's scapular area where a reflex "locking" action occurs, holding beamer's arms and therefore also the indicator in an unnatural off-center position.

It is the difference in a beamer's reflex intensity (BRI)—between 0% BRI (on-course), Figure 1a, and 100% BRI (off-course), Figure 1b—that affects the directional on-course guidance factor.

beamer's reflex action diminishes, and the indicator tends to fall to a centered (natural) position. It is the BRI differential between off-course and on-course orientation of the beamer that induces the subtle reflex of directional beam guidance.

Biogeomagnetic reflex action in the human animal, although inherent, lies in a latent state through lack of use. For this reason, initial attempts at proper biobeaming procedure may result in some disappointment to a beamer expecting immediate, consistent results.

Beamer reflex action is as natural as those reflexes that constantly adjust our balance against the forces of gravity. This action possesses subtle characteristics. A maximum degree of accuracy and consistency can be accomplished only through patient, determined, repeated practice in allowing the beamer's arms to loosely swing at the shoulder joints in a completely relaxed manner while all other beaming procedure is observed.

Initial attempts at detecting a beamer's reflex action should always include use of the beaming indicator as shown in Chapter 4. The indicator magnifies the degree of detected BRI, a very important tool for fledgling beamers.

Once a beamer can visually detect any degree of off-center beaming indicator positioning, the following quick, simple test will verify credibility of this reaction:

> Repeat same procedure, gently recocking indicator to left or right (depending on which side has been chosen for this beaming procedure). If an off-center position of indicator persists while continuing the procedure, shut eyes tightly to eliminate all light. The beamer will then immedi-

ately sense the indicator as dropping to an unlocked, on-center (0% BRI) position. A witness can also make this observation.[3]

To best describe beaming guidance, it might be said that the on-course reaction in biogeomagnetic beaming detection is simply a matter of following the line of least resistance. This "line" is the lateral direction in which the bineural guiding reflexes are most at rest (i.e., 0% BRI). Such a reaction is referred to as *lateral, directional equilibrium.*

That familiar form of involuntary directional guidance that keeps us on an even keel with the force of gravity has been long recognized. Considering that nature provides animals with such reflex mechanisms for assured vertical equilibrium, common reasoning dictates that another dimension in the form of lateral, directional equilibrium would not be omitted. Without both, many animals could not survive.

The most difficult problem to overcome during biobeaming procedure is recognition and control of one's subconscious influence on certain otherwise unhampered reflexes. It is a constant threat to the biobeaming accuracy of all human beamers, particularly the inexperienced. There is, however, a rule-of-thumb to remember:

[3] To intentionally belabor the point, only through frequent practice will those long- neglected biobeaming reflexes develop in efficiency for accurate lateral directional beaming results. Only by persistent practice, punctuated by errors, does a young child, in combating the vertical force of gravity, gradually develop specific reflexes enabling eventual stability in maintaining a vertical stance.

- Biobeaming involves only *one* basic reflex action. It is the off-center, locked position of a beamer's arms and hand-held indicator (positive reaction). (See Figure 1b.)
- Contrary to this reflex action is *no* reflex at all when a beamer's arms and indicator gravitate to a loose, neutral position aligned with the vertical center of the beamer's body (negative reaction). (See Figure 1a.)

A beamer soon learns to sense whether the indicator is locked in the off-center position or is being subconsciously held in that unnatural alignment against gravitational force. However, the greatest danger of error lies in beaming reflex action when the hand-held indicator may be subconsciously induced to fall to a 0% BRI alignment with the center line of a beamer's body.

Carelessness during the beaming procedure has proven to be the enemy. Only careful recocking of the indicator during beaming procedure retests will assure accuracy during early biobeaming practice. Once one fully realizes the existence of biobeaming reflex phenomena, the incentive will be strong for curious individuals to further develop this ability. It will, in addition, serve as a catalyst to inspire mounting interest in overall geomagnetic directional beaming phenomena.

Determining the beam direction and intensity

Following the beaming indicator procedure, one will learn to detect if a directional beam is in force, its lateral orientation, and its intensity.

Listed below are the necessary terms, definitions, and instructions involved in detecting directional beams through biobeaming reflex action.

- The term *beam,* when applied in this study, is defined as the disrupting geomagnetic influence found to affect a beamer's specific motor neuro reflexes. This phenomenon is detected when a beamer faces in any but *one* given lateral direction within a 360-degree radius.
- The result of this *beam* upon a beamer is a subtly induced neuro reflex deflection toward that *one* specific direction.
  - When in this one direction, *no* disrupting neuro reflex deflection exists. This is referred to as a beamer's *on-course orientation.*
  - All other directions faced by a beamer within a 360-degree radius are referred to as a beamer's *off-course orientation.*

This directional beaming action is referred to as the beamer's *reflex guidance factor.* Theoretically, the beamer's reflex guidance factor subconsciously guides an insecure bird, with absolutely no sense of navigation, in a given direction. The animal is fully unaware that it is homing or on a migration course. It finds solace through flight deflection in that one direction, effecting a secure state of lateral directional equilibrium through the absence of any reflex disruption.

- *Beamer reflex intensity (BRI)* is the detected and calibrated degree of a beamer's induced directional beam influence, ranging from a maximum 100% BRI to a minimum 0% BRI.

- *BRI differential* is the resulting differential intensity between a beamer's greater off-course BRI and the lesser on-course BRI. Under normal circumstances, these BRI differentials follow a pattern.
  - During migration seasons (fall and spring), a 100% BRI off-course reaction during either migration or homing will result in a 0% BRI during all on-course reactions, resulting in a 100% differential.
  - During the nonmigration seasons (summer and winter), the all-season migration beam induces 50% BRI in all lateral directions faced. The all-season homing beam induces only a 75% off-course BRI. The on-course BRI, however, remains at 0%.
- *No directional beam (detection)* refers to circumstances when a beamer's BRI is the same (normally 50%) in all directions faced.
- *No beam (detection)* refers to total loss of all biobeaming reflex (0% BRI) in all directions faced (e.g., beaming tests conducted with the eyes of the beamer tightly closed, or tests conducted in close proximity to known beam disrupting forces). (See Chapter 5.)
- The term *faced by beamer* means any lateral direction aligned with the forward binocular vision of a beamer, and perpendicular to the broadside of a beaming indicator, both serving as directional antennas.

BRI variations

The biobeaming phenomenon is complex; the biobeaming procedure is not. Nevertheless, when developing the latent biobeaming reflexes through practice in proper procedure, some pitfalls lie in the path of progress. The following problems have been found to cause detected irregularities in expected beaming results.

- A 0% BRI reaction in all directions faced under normal beaming procedure is nonexistent. It can, however, be induced by one or more of the following:
  - Poorly designed or improperly held beaming indicator
  - Beamer's procedure with eyes tightly closed or when beaming attempted in *total* darkness
  - Beamer wearing Polaroid lenses during tests
  - Beaming procedure in proximity of other beam disrupting forces
- Biobeaming procedure, which requires the beamer to turn laterally in search of a possible 0% BRI (on-course) direction, poses a potential problem for a beamer intent on conducting beaming tests. To prevent this:
  - While properly holding beaming indicator, avoid a 360-degree counterclockwise body rotation at all times in searching out a potential directional on-course orientation. This precaution is necessary, since three full counterclockwise body rotations establish a homing

beam source on the spot or area of the rotations. While less than three rotations will not fully accomplish this, it will temporarily disrupt the area for further beaming tests.

- Also, *during* counterclockwise body rotation under the above conditions, BRI readings will tend to fall toward zero. Any such loss of BRI during body rotation is a warning that a homing beam is being established.

Proper beaming procedure is the key to achieving biogeomagnetic reflex action and learning to detect directional beams and their characteristics. Accurate results cannot be assured without a full understanding of the techniques involved.

3 Directional Beams

While investigating this newly discovered beaming phenomenon, the following three individual directional beam forces were discovered (listed in sequence of their discovery):

1. The migration beam
2. The homing beam
3. The shoreline (or water) beam

The guiding direction of each beam is always indicated by the *decreased* degree of the beamer's reflex intensity (BRI) when directed toward that beam source (the on-course direction).

This chapter describes *directional beams,* based on accumulated data from studies conducted in the northern hemisphere under normal, seasonal geomagnetic conditions.[1]

---

[1] All beaming tests were conducted in the northern hemisphere. The purpose for stating this fact is based upon certain test results when using a bar magnet in proximity of these tests. Such test results repeatedly suggest the possibility that the seasonal migration beam orientation in the southern hemisphere may veer 90 degrees from that direction detected in the northern hemisphere during the same season.

The migration beam (northern hemisphere)

The migration beam, first of the three directional beams to be discovered, is recognized in the northern hemisphere by the following specific characteristics:[2]

- Between the March equinox and June solstice, this on-course directional influence lies toward the *north-west.*
- Between the September equinox and December solstice, this beam's influence shifts 180 degrees, so it is oriented with its directional on-course influence toward the *southeast.*
- Between the solstices and equinoxes, *no* migration beam reflex action can be detected.
- Figure 2 depicts Earth and the seasonal migration beam orientations. This illustration shows a 90-degree shift in seasonal migration direction at the

---

[1] continued:

1. The northern hemisphere fall migration beam, detected to the southeast, is suspected to veer to the southwest below the equator.
2. The northern hemisphere spring migration beam, detected to the north-west, is suspected to veer to the northeast below the equator.

[2] A large percentage of migration beam tests were conducted in Twenty-nine Palms, California. Other tests in areas within a 1,000 mile radius indicate similar northwestern and southeastern orientation results. Because of the lack of more widespread longitudinal testing, it is yet to be determined, but it is suspected, that seasonal migration beam influence can exist at any longitude. It would, however, be subject to lateral deviation in orientation relative to the degrees of polar magnetic field deviation at given longitudes.

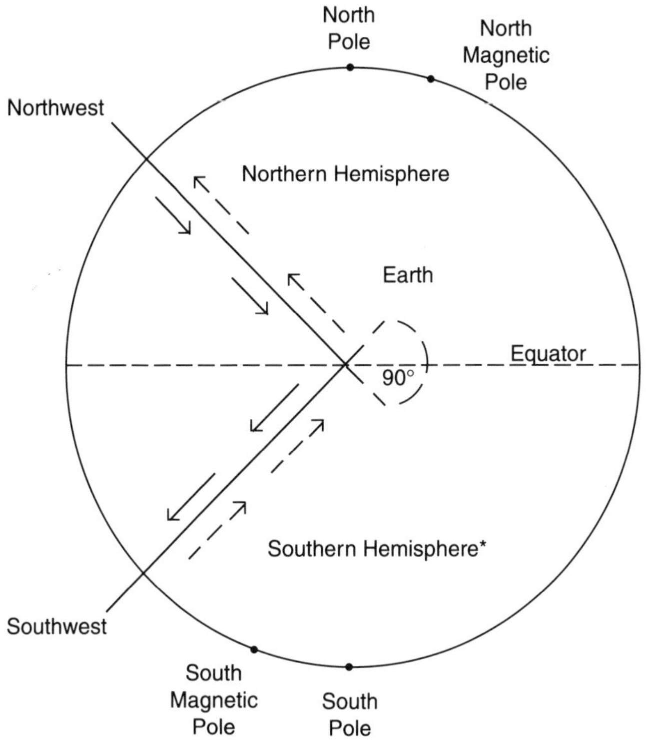

* Southern hemisphere data based upon test results from LMR apparatus tests. To date, no actual on-the-spot tests have been made in the southern equatorial hemisphere.

——► Indicates influential direction of seasonal migration beam orientation in the northern and southern hemispheres between the September equinox and the December solstice (all latitudes and longitudes)

– – ► Indicates influential direction of seasonal migration beam orientation in the northern and southern hemispheres between the March equinox and the June solstice (at all degrees of latitude and longitude)

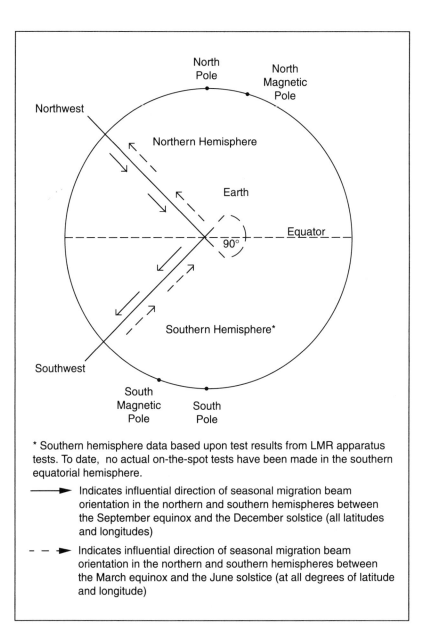

Figure 2. Seasonal migration beam.

equator as has been repeatedly found to exist in simulated LMR apparatus tests.

- The seasonal migration beams have only one characteristic in common with the magnetic compass, that is, migration beam detection, when in force, is believed to be possible in all longitudes and latitudes in either the southern or northern hemispheres.
- The migration beam, when in force, is not affected by massive obstructions, such as mountains or tall buildings, standing in its path.

After discovering the existence of the forces responsible for the seasonal migration beam, the question arose as to what existing conditions might trigger this phenomenon. Suspecting the Earth's polar magnetic field to be a factor, the approach was to conduct beaming tests in proximity of a bar magnet as an apparent simulator of that geomagnetic force. These tests were conducted during a period of seasonal migration beam influence. The following was discovered:

- Placing a bar magnet in natural orientation with Earth's polar magnetic field produced no disruption of a prevailing seasonal migration beam.
- With the magnet placed in an opposing orientation with Earth's magnetic field, a magnetic disruption occurred canceling all prior detection of that same seasonal migration beam. The beamer's reflex action was then reduced to one of 50% BRI in all lateral directions faced, negating all directional beaming.

Here, then, was a discovery apparently relating Earth's polar magnetic field with the beaming phenomenon. Other factors, however, demanded investigation.

Well aware from repeated research that a beamer's retinal light reception is a critical factor in inducing beaming reflex action, and considering, as well, the electromagnetic (light) properties involved in the Zeeman/Faraday Effect, experiments were made applying the *light dominance* factor to the already revealing bar magnet experiments.

In preliminary experiments, the north-seeking pole of a bar magnet was coated white and the south-seeking pole was coated black. Figure 3 illustrates the variations in bar magnet design used in magnetic field/light dominance beaming test procedure. These white/black applications converged at the magnet's neutral center (equator) line. With this altered bar magnet, additional beaming tests were conducted. The findings of this preliminary investigation strongly reinforced the indications of an existing relation between electromagnetic (light) forces, geomagnetic forces, and biobeaming phenomena. This apparent breakthrough led to an extended series of studies referred to as the *polar light dominance factor*.

The polar light dominance factor

In further investigating this initial light dominance experiment, other results emerged. It was found that a naturally oriented bar magnet with a white north-seeking pole and a black south-seeking pole greatly altered the outcome of those prior tests. The magnet in natural geomagnetic field orientation triggered the following results:

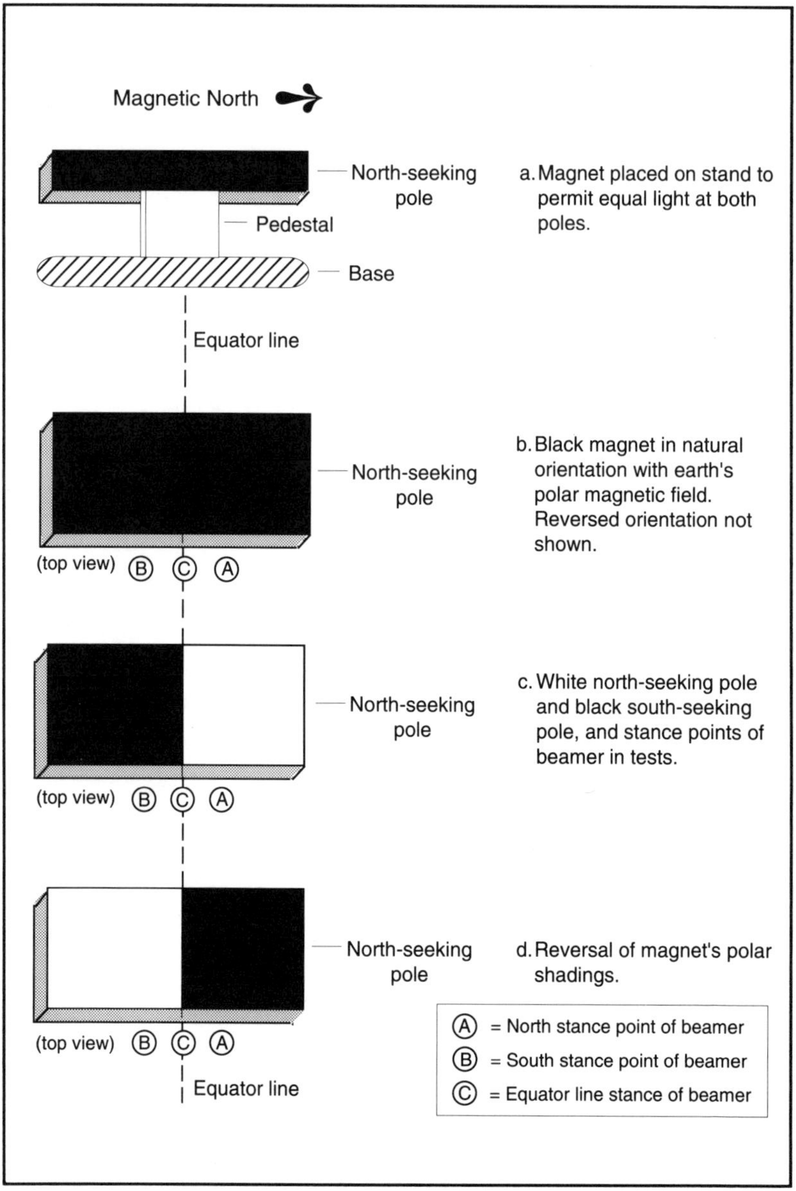

Figure 3. Bar magnet ($1\frac{1}{4}$" x $\frac{3}{4}$" x $\frac{1}{2}$").

- When a beamer stood *north* of the magnet's neutral line (see Figure 3), an influencing on-course northwest directional beam reaction could be detected.
- When a beamer stood *south* of that "equator line," the migration beam influence continues to exist—but *not* to the northwest. The direction of on-course influence is then to the northeast. A 90-degree shift in beam direction occurs when a beamer crosses the bar magnet's equator line (see Figure 3).
- When a beamer stands aligned with the line of the bar magnet's neutral equator (magnetic east or west of the bar magnet's center point), no beaming reflex action (0% BRI) is possible.

In further investigating this phenomenon, the light dominance factor, applied to the magnet's poles, was then reversed with the north-seeking pole being *black* and the south-seeking pole *white*. The repeated results from this experiment were no less surprising. When reversing the light dominance factor at the magnetic poles, the orientation of the seasonal migration beam was also reversed.[3]

After years of testing, these consistent discoveries were the factors responsible in first forming the concept that migration beam direction in the *southern* hemisphere is a reversed mirrored reflection of that in the northern hemisphere in any given spring or fall season.

---

[3] The term *light dominance* refers to light intensity differentials. In these early tests the extreme differentials (black/white) were used. Later notes will show that light intensities of lesser differential produce the same effect.

The discoveries relating the beaming phenomenon to a magnetic fields interaction with polar light conditions were the forerunner to designing the LMR directional beaming simulator apparatus. (See Chapter 5.) The added factor of lateral *rotation* action in LMR design made the controlled beam simulation procedure a most valuable tool. Simulation of seasonal migration beam conditions in any season were then instantaneous, eliminating prior need to await the arrival of a given season in order to conduct a given test.

When first relating the normal, seasonal biogeomagnetic migration beam to the known patterns of bird migrations, for example, it was apparent that bird migration patterns reflected the spring and fall reversals of the beam orientations (Figure 2) and the absence of migration beam detection during the winter and summer seasons. Later biobeaming studies revealed the existence of a homing beam and a shoreline beam emphasizing that biobeaming phenomena was indeed complex. With this discovery, however, the long-known mysteries of animal homing as well as migration feats appeared to diminish proportionately.

The homing beam

The homing beam phenomenon was inadvertently discovered in the course of migration beam investigations. The revelation of this second beam force was understandable due to the conditions that trigger its existence. Body rotation (or circling) is a normal gyration when searching for a desirable direction of travel. The beamer's body maneuvers involuntarily establish a homing beam source. The homing beam and its northern hemispheric procedure are as follows.

Using the normal indicator-holding procedure, a homing beam source is established through three counterclockwise body rotations on a given spot (or circling of a given area).

- An initial 360-degree counterclockwise body rotation[4] will result in a total loss of beamer's reflex intensity (0% BRI reflex action in all directions faced). This first rotation produces only an unstable homing beam source.
- A second such rotation results in a now disruptive full 100% BRI in all directions faced by the beamer from that rotation or circled area.
- The third 360-degree rotation, however, on the same area will result in a stable (0% BRI), permanent homing beam source.

Additional body rotations on this spot or area circled will no longer affect this now established, stabilized homing beam source.

The characteristics of a stable homing beam source are as follows:

- A 0% (on-course) reaction occurs while the beamer stands on the established spot or within the circled area, regardless of the direction laterally faced by the beamer. Figure 4 illustrates any degrees of indicator calibration "stop" position between 0

[4] A relaxed, no-reflex action (0% BRI) occurs only during a beamer's counterclockwise direction of rotation. This action induces within an insecure, lost bird (for example) the tendency to circle in that direction, ensuring the establishment of a homing beam source.

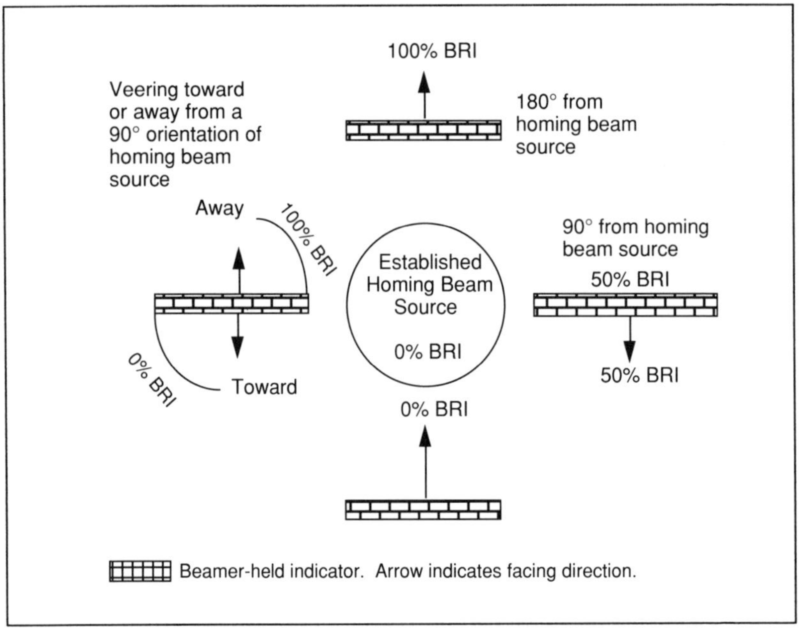

Figure 4. Homing beam.

degrees (on-course) and 180 degrees (off- course) that reflect the beamer's facing orientation during the same period. The beamer's reflex intensities (BRI) occur when maximum BRI potentials exist.

- From any other area, the 0% BRI (on-course) reaction occurs only when the beamer's body is oriented with that homing beam source while the beaming indicator is simultaneously positioned broadside to the same direction.
- The direction where beam reflex action registers 100% maximum BRI (off-course) is an orientation that is 180 degrees from the 0% BRI (on-course) direction.

- A beamer's reaction to the homing beam source parallels that of migration beam detection. In the process of turning toward an established homing beam source (either clockwise or counterclockwise) when within 90 degrees of that source, the degree of beam reaction falls to 0% BRI. Figure 5 shows how an established homing beam source, when within effective range of a beamer of equal color-value status, can predominate over a beamer's detection of an existing seasonal migration beam.
- The lesser BRI reaction, when on-course, will immediately result in a greater BRI (off-course) reaction if a beamer deviates from that direction either clockwise or counterclockwise. Only when again veering or turning toward the homing beam source will the lesser BRI (on-course) reaction be reinstated.
- Effective homing beam sources of differing color-value frequencies can be established at any point where other established homing beam sources exist. This does not disrupt the effectiveness of prior established sources.
- A homing beam source can be thought of as a form of frequency modulation antenna. This conclusion was reached following consistent results of repeated self-biobeaming research tests. The established homing beam antenna is found to emit its beam influence 360 degrees laterally from a verti-

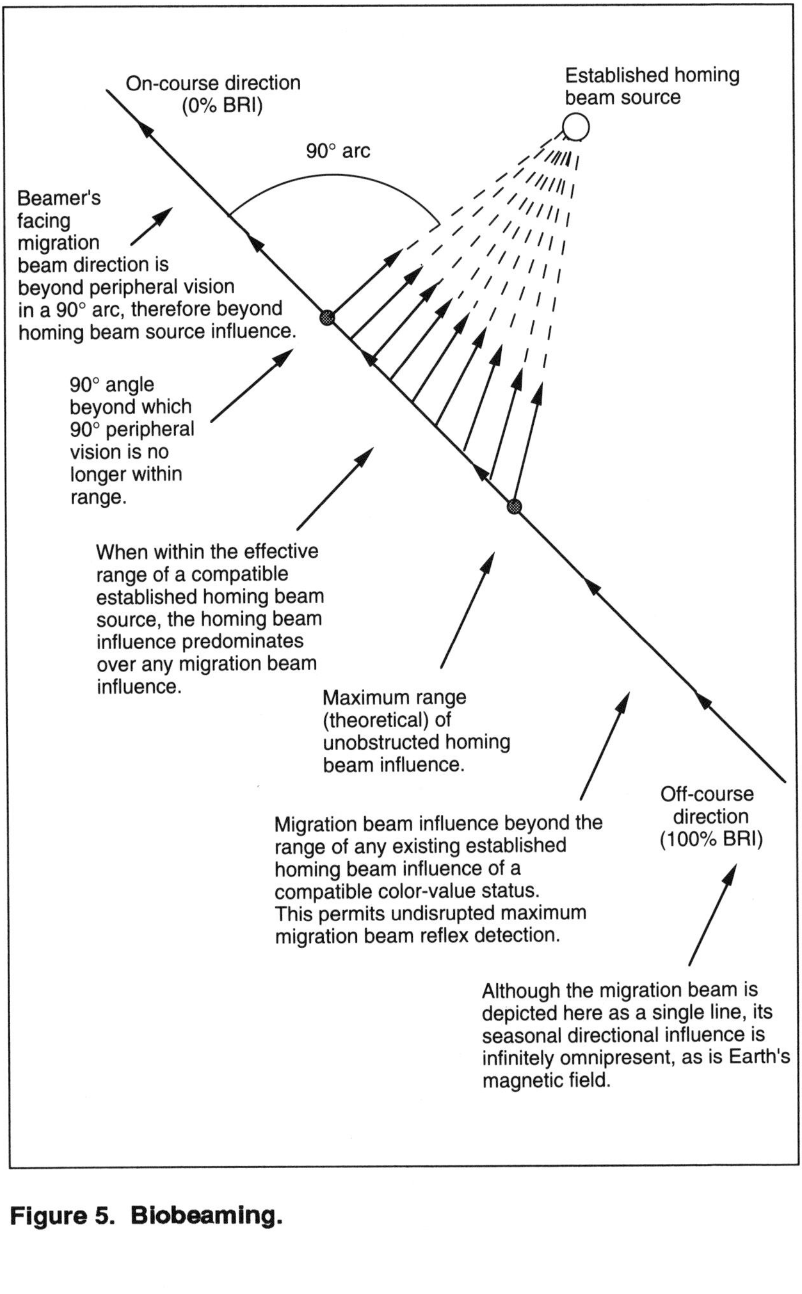

Figure 5. Biobeaming.

cally established homing beam source. Such antenna height is not established to date, but limited testing conditions indicate this height potential to be well within requirements for all animal homing efficiency.

- Detection of these lateral homing beam emissions are found to lie on a plane between horizontal and 45 degrees below horizontal. The effective range is affected by Earth's curvature or heavy mass obstructions such as mountains or tall buildings. Figure 6 illustrates the theorized range of homing beam emissions and the resultant distance of influence determined by high obstructions (in the figure, mountains) or Earth's curvature.
- All directional beaming, including homing beam detection, is *both* nocturnal and diurnal. Unlike the seasonal migration beam, homing beam source establishing and detection has a potential in all seasons, making off-season migrations possible.
- No additional stable homing beam source can be established within the influencing range of a prior established source of *like* color-value status. The maximum stability of an established homing beam source has not been determined to date. However, such an established source (used for beaming test data) has endured in excess of two years with no indication of deterioration. (For *intentional* cancel-

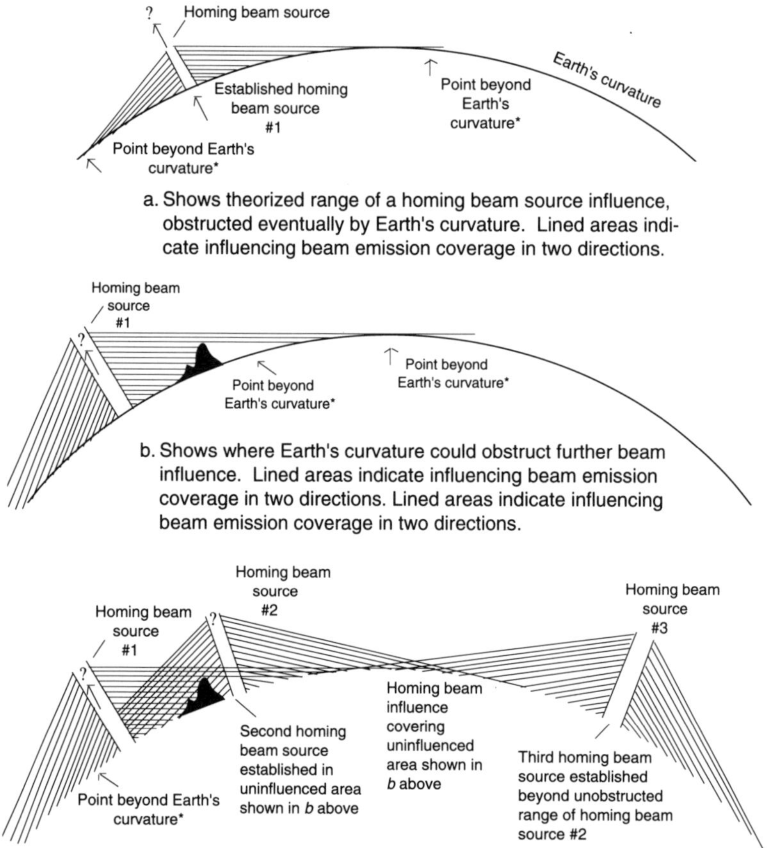

a. Shows theorized range of a homing beam source influence, obstructed eventually by Earth's curvature. Lined areas indicate influencing beam emission coverage in two directions.

b. Shows where Earth's curvature could obstruct further beam influence. Lined areas indicate influencing beam emission coverage in two directions. Lined areas indicate influencing beam emission coverage in two directions.

c. Shows how other primary homing beam sources could eventually be established in those uninfluenced areas due to obstructions of homing beam #1, evolving into a continuous path of homing beam guidance between the three distant points shown. Lined areas indicate influencing beam emission coverage in two directions.

*Point of established homing beam sources. Maximum height is undetermined.

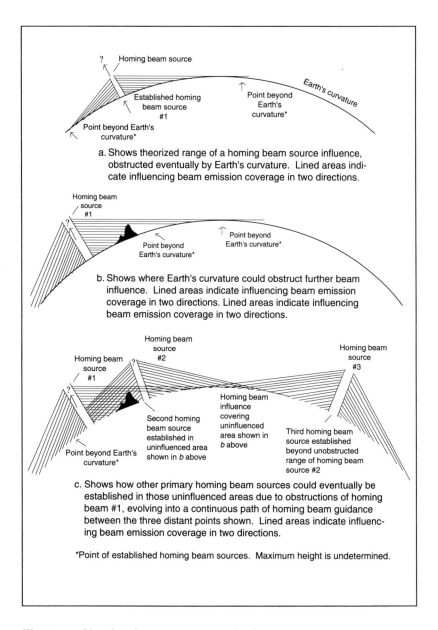

Figure 6. Homing beam source emissions.

lation of a homing beam source, see Chapter 5, "Pivot-to-Ground Factor.")

- Tests repeatedly show that homing beam sources established in or above a body of water are not compatible with those established over land areas, and vice-versa. Figure 7 illustrates the theoretical development of a migratory homing beam flyway over a water or land route with no shoreline or seasonal migration beam influence as added induced guidance. The aggregate of circles illustrates homing beam sources established in early development of scattered homing beam paths between two points. In early development, the aggregate circles (i.e., established primary homing beam sources) reflected wanderings of birds, which developed a network of overlapping homing beam sources in a reasonable general direction of future safe and fruitful migration. In the course of time, such a network of homing beam influence developed into a guiding migration beam pathway and eventually evolved into an established migration flyway (solid arrows) following the most safe, successful route due to the overall distance factor.

The involuntary establishment of homing beam sources has the potential of affecting an important link in directional guidance of migrating or homing animals. Based upon extensive homing beam test study, such homing pathways could evolve from the normal gyrations of a bird. For example, it would be a natural

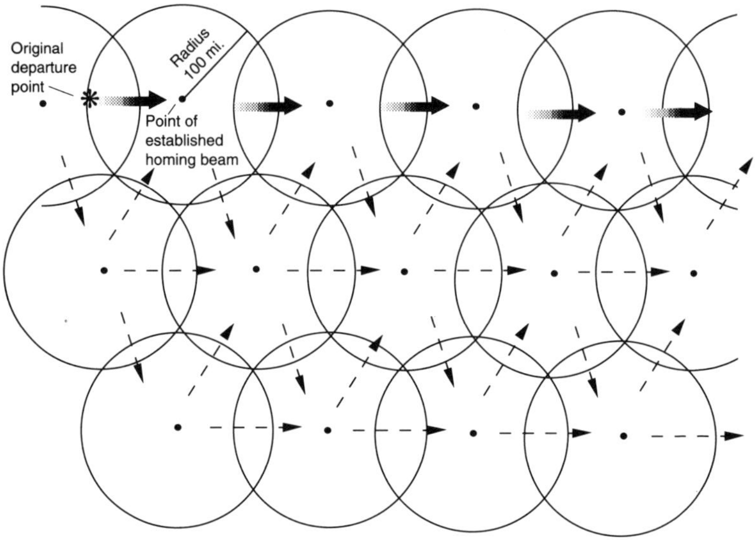

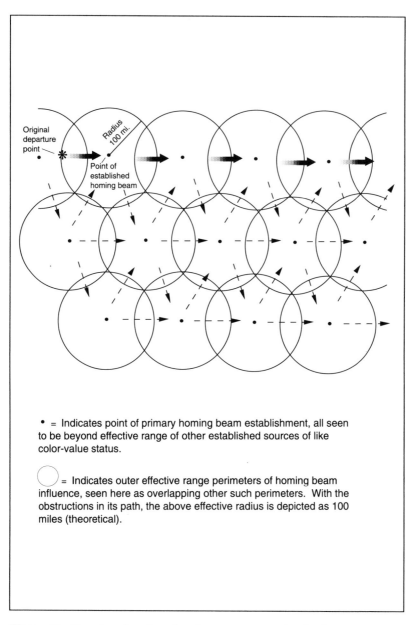

• = Indicates point of primary homing beam establishment, all seen to be beyond effective range of other established sources of like color-value status.

 = Indicates outer effective range perimeters of homing beam influence, seen here as overlapping other such perimeters. With the obstructions in its path, the above effective radius is depicted as 100 miles (theoretical).

Figure 7. Overlapping homing beam sources/emissions.

reflex to circle an area when the urge to take flight is accompanied by no existing sense of intended direction.

As illustrated in Figure 7, homing beam sources established at any point beyond the range of an existing homing beam of like color-value status (discussed later), yet near enough to affect the overlapping of the two homing beam maximum ranges, would link the two into one continuous source of homing guidance. Such repeated occurrences would eventually develop into courses of homing beam directional influence between two points. Through repeated usage, the route that most effectively guides the birds toward their unknown but predestined goal would eventually evolve into an established migratory flyway, just as heavily traveled routes of land animals are reflected by the well-beaten land trails they follow.

The fact that separate bird species do, however, branch out from the general flyways in selective directions toward the end of a migration emphasizes the potential of the color-value factor, described in the following section.

Color-value factor

As explained earlier, the effects of light and its interaction with other known forces produce the biogeomagnetic directional beams. The element referred to here as *color-value* is a critical facet of the homing beam phenomenon. It is a process where light interaction assumes an additional role.

In biobeaming procedure, a beamer cannot detect the presence of any directional beams until such light is in contact with both retinas of the individual. Where the homing beam is involved, however, these electromagnetic light waves take on an added dimension in the form of an invisible, infinite color spectrum. It is this wide range of color and its values that are responsible for the selective homing capabilities of many homing species.

A discovered integral part of the homing beam phenomenon, identified as the color-value factor, is theorized to be the key to selective homing guidance, particularly during animal migration periods. The color-value factor renders a given established homing beam source selective to the established and all other beamers of identical color-value status. To clarify the term *color value* as used to identify this phenomenon, the color (hue), alone, is not the sole influencing factor. The value of the color is equally involved. Color-value tests, using varied shades of the achromatic scale (e.g., black to grays to white) are equally effective in homing beam detection. Repeated tests have produced consistent results showing that the color-value factor must be carefully considered to ensure repeatedly stable and compatible response to a specifically established color-value homing beam source.

Test results show that a homing beam source, established by a beamer, can induce a homing reaction only if the beamer does not alter the influencing color value existing at the time the homing beam source was established. The homing beam source retains attuned synchronization of the color-value wavelengths between the beamer and the established homing beam source. Such findings necessitated a clear understanding as to exactly what factors determine the color value of a given beamer. Homing beam/color-value experiments established the following facts:

- The end (combined) color value that directly contacts[5] and spans the center line of a beamer's forward body side is the influencing color-value factor.
- All influencing color values—their hues, colors, and volume—when combined, determine the end color value (i.e., the beamer's color-value status). Influencing natural color-value factors would include flesh tones and hair (plumage and beak, for a bird).
- The alteration of any influencing color value will change the end color value of the beamer, much as the addition of any given pigment color in paint would change its end color-value status.
- With humans, the beaming indicator, used as an external tool and held properly, spans the center line of the body and directly links the left and right sides of the bineural system. In so doing, the indicator becomes a factor in the beamer's color-value status (e.g., an indicator painted white would contribute the value of white to the overall color-value status).

[5] Past notes verify the time spent with the color-value factor covering this facet. Such a discovery was prompted by the puzzling fact that a beamer's change of outer clothing does not affect color-value status in homing procedure. The attempt to understand this problem emerges as one of the most frustrating facets of biobeaming exploration. Not until the factor of color-value body *adherence* was considered did the answer become apparent.

> If any color or differing value is then added so as to cross the center line of this indicator, such added color and/or values would again alter the end color-value status.

The findings involving the color-value factor have been repeatedly reinvestigated throughout the years, producing the same consistent results. The next section on color value shows how the color-value factor would apply during bird homing activity.

Color value in relation to bird homing and migration

Following the discovery of the migration beam and homing beam, the realization slowly developed that the homing beam with its fine-tuning color-value facet could play a major role, not only in relatively short-distance homing, but in the long-distance migration of some animals, birds being a prime example. In relating the color-value factor to animal homing and migration, one observes that many birds undergo seasonal alterations in their color-value status. Land birds, in general, sport brighter plumage during the spring and the fall migrations. This would in no way hamper their seasonal migrations as each species would react to the established homing beam sources of their predecessors in both of these seasons. If some variance existed between the migrator's spring and fall color-value tones, separate homing beam wavelengths for separate seasonal directional flights would assure proper one-way guidance for each direction. Such homing sources, inadvertently established by circling or body rotation, dot migration paths with beaming "beacons" of long duration and potential range, directing varied species of birds, for instance, to the habitual feeding/nesting areas of their forebears.

Working in conjunction with the prevailing seasonal migration beam, which provides guidance in a general direction at any longitude and latitude, the homing beam improves on such directional influence by funnelling the migrators along specific flyways established by their predecessors, providing a near-constant added protection against wind drift during migration beam travel and more assurance of water and feeding grounds while en route. Based on homing beam tests, the possibilities are strong that a migrating bird is deflected from a given migration beam course to a compatible color-value homing beam course within range due to the specific characteristics of the homing beam. The relaxed on-course reflex action induced by the migration beam would converge with a similar influencing homing beam source either left or right of the bird's migration beam direction. The on-course direction would then seem to split or broaden in the direction of the now influencing homing beam. Once a traveler, under migration beam influence, veers slightly from the migration beam course and within a 90-degree arc toward the beckoning homing beam source, it would be strongly guided toward the latter direction. (See Figure 5.)

Based upon the studied characteristics of biobeaming, now experienced by the reasoning human, one factor is apparent. The guiding influence is not voluntary, but rather an induced biological reflex of lateral directional equilibrium. In the same manner that water seeks its own level, the deflection from the migration beam influence toward the homing beam source parallels that of a stream of water seeking the most downhill course—the path of least resistance.

It would appear that the seasonal migration beam and homing beam influences would suffice for effective migrational guidance. However, nature offers yet another dimension of protection, as the discovery of the shoreline beam will indicate.

Shoreline beam

The third and final directional beam to be studied, the shoreline beam, adds yet another dimension to the overall network of discovered geomagnetic directional guiding forces. In the course of investigation, it was repeatedly evident that beaming results were often disrupted when tests were conducted near water. More concentrated study of this phenomenon led to a discovery referred to as the *shoreline beam* (water beam).[6]

Tests repeatedly verified that the shoreline beam possesses the following unique characteristics not associated with the migration beam or homing beam.

- The source of the shoreline beam originates at the water's edge. From this "shoreline" the beam forces are detected as apparently fanning out 180 degrees at all points perpendicular to the shoreline, covering

[6] It is not intended here to relate such water-detection phenomena to those claims of others referred to as *dowsing* (water witching). This statement is based upon the fact that in no biobeaming procedure experiences has water detection been possible that suggests presence of water deposits at the more excessive depths of average underground water tables.

angle planes from horizontal to vertical. (See Figure 8.)

- As with the detection of the migration and homing beams, the broadside-held position of the curved or vee-shaped indicator and the beamer's body align-

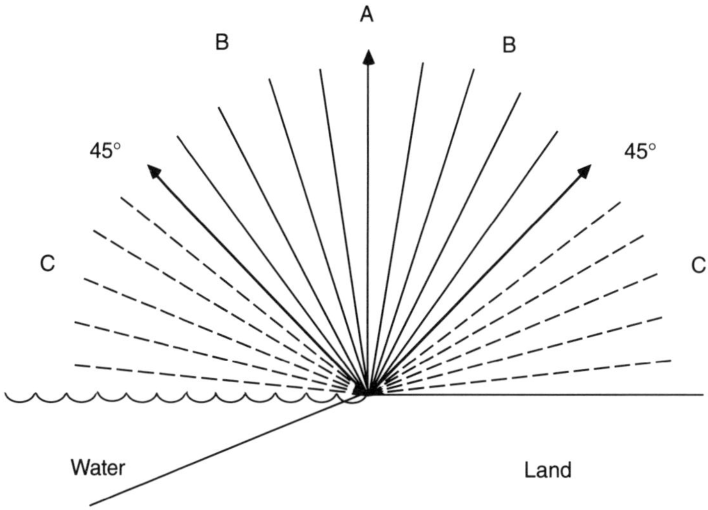

A = Emanating lines fanning out from the shoreline, representing shoreline beam forces, the source of the shoreline beam.

B = In shoreline beam detection procedure, at solid line area shown, the beaming indicator arc may be directed downward (vertically) or toward the shoreline.

C = In shoreline beam detection procedure, at broken line area shown (below 45° plane) beaming indicator arc must be directed toward shoreline only.

The top view of this illustration (not shown) would simply depict all beam forces as emanating perpendicularly from all points along the shoreline

Figure 8. Cross-section of shoreline beam.

ment facing the direction beam source become the determining factors.

- When approaching a shoreline, and when the beamer is positioned with the broadside of the indicator within a nearly horizontal- to a 44-degree angle above the water's edge (Figure 8c), no degree of on-course reaction will be detected.
- When the position of an indicator, relative to the point at water's edge, lies at a plane of approximately 45 to 90 degrees above the shoreline:
  - An on-course reaction is detected with the indicator held vertically or at any tilt between downward (vertically) and a 45-degree angle from that position.
  - The beaming indicator, when tilted downward toward the water's edge, triggers a full on-course (0% BRI) reflex action. This occurs at any angle plane from the shoreline lying between 45 degrees and directly above that shoreline.
  - The vertically held indicator, in comparison, induces only a weak on-course BRI reaction within the beamer when tested at the 45-degree angle above the water's edge. This weak on-course reaction becomes more pronounced the closer it acts to the water's edge. The full on-course (0% BRI) is reached when the indicator is held directly above that shoreline (water's edge). (See Figure 8b.)

- With the beamer directly above the water's edge, a change occurs along that shoreline. The beam now assumes seasonal, lateral directional characteristics in conjunction with any current seasonal directional migration beam orientation. The following was found:
  - With the shoreline directly below the beamer, on-course detection occurs when the beamer is oriented so that a vertically held indicator's broadside faces parallel with the shoreline orientation.
  - At this point, the guiding direction of the shoreline beam is determined by the prevailing season, in that between the solstices and equinoxes (the seasons when a migration beam is not detectable), the shoreline beam is found to be bidirectional. This condition then induces an on-course reaction when the beamer faces in either direction along that shoreline. Figure 9 illustrates river and shorelines as an example of a shoreline directional beam source in the northern hemisphere. The broken arrows show the direction of the spring shoreline on-course influence. The solid arrows show the direction of the fall shoreline on-course influence. The solid two-way arrows show the bidirectional orientation of on-course influence during the winter and summer seasons. All of the seasonal directions shown in Figure 9 are with the beamer directly above the shoreline. The west-east shoreline illustrates that no directional shoreline beam

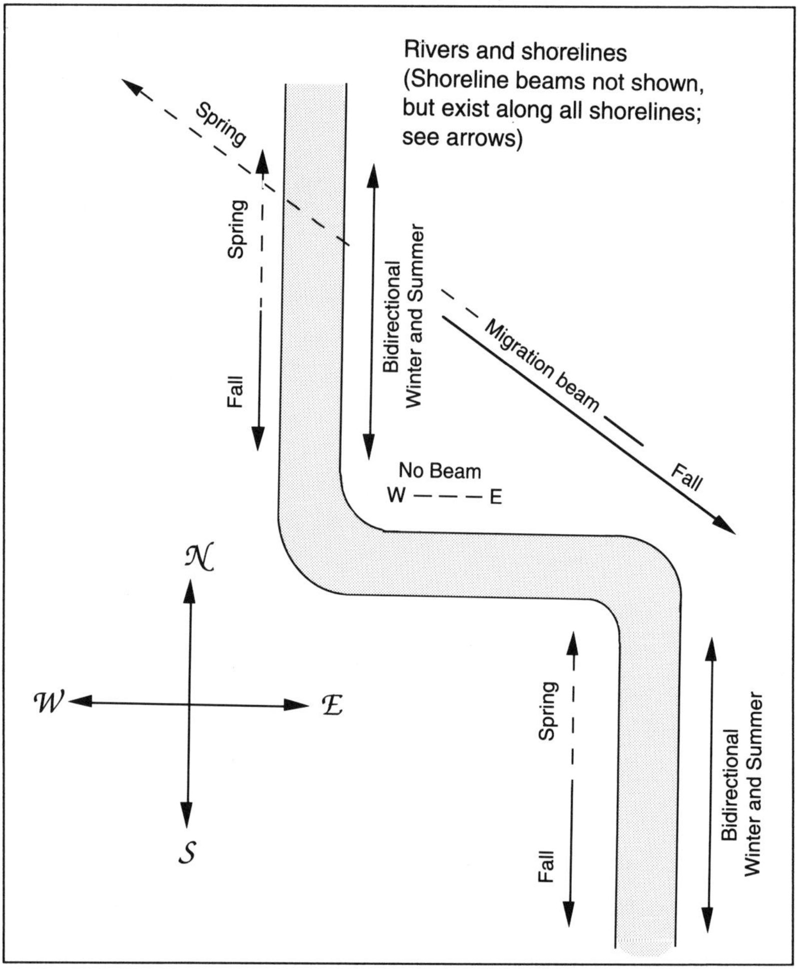

Figure 9. Shoreline beam in relation to migration beam orientation

can be detected, regardless of the prevailing season (no 0% BRI differential).

- Between the equinoxes and solstices (the migration beam seasons), the northern hemisphere shoreline beam is found to be unidirectional. The on-course direction is now determined by the prevailing migration beam direction. With a seasonal migration beam guiding influence now to the northwest, a shoreline running at any northward degree will induce a shoreline beam reaction along that shoreline in that direction only. During the fall, the northern hemisphere migration beam reverses to the southeast, dictating a reversal of the shoreline beam orientation and inducing a directional beam guiding reflex southward along that shoreline.
- When directly above a shoreline (on-course) any tendency to turn or veer either to the left or right induces a strong degree (100% BRI) off-course reaction. Only by correcting this directional drift will the full (0% BRI) on-course reflex reaction be reinstated.
- Under those circumstances, however, when a compatible land-established homing beam source lies within range and within a 90-degree arc of a beamer on a shoreline course, the drift factor is altered. Any tendency to veer toward land will not interrupt an on-course reaction. Thus, the homing

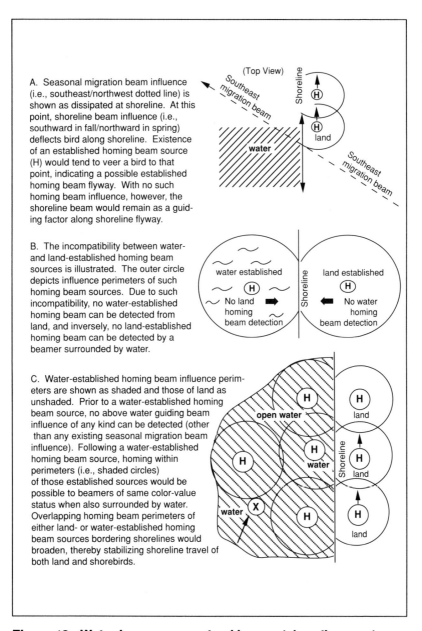

A. Seasonal migration beam influence (i.e., southeast/northwest dotted line) is shown as dissipated at shoreline. At this point, shoreline beam influence (i.e., southward in fall/northward in spring) deflects bird along shoreline. Existence of an established homing beam source (H) would tend to veer a bird to that point, indicating a possible established homing beam flyway. With no such homing beam influence, however, the shoreline beam would remain as a guiding factor along shoreline flyway.

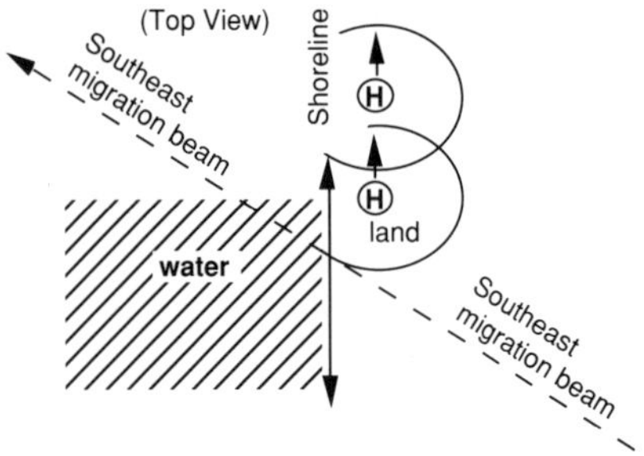

B. The incompatibility between water- and land-established homing beam sources is illustrated. The outer circle depicts influence perimeters of such homing beam sources. Due to such incompatibility, no water-established homing beam can be detected from land, and inversely, no land-established homing beam can be detected by a beamer surrounded by water.

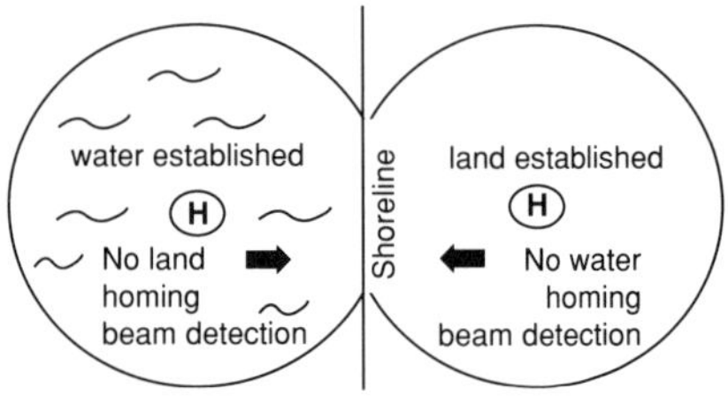

C. Water-established homing beam influence perimeters are shown as shaded and those of land as unshaded. Prior to a water-established homing beam source, no above water guiding beam influence of any kind can be detected (other than any existing seasonal migration beam influence). Following a water-established homing beam source, homing within perimeters (i.e., shaded circles) of those established sources would be possible to beamers of same color-value status when also surrounded by water. Overlapping homing beam perimeters of either land- or water-established homing beam sources bordering shorelines would broaden, thereby stabilizing shoreline travel of both land and shorebirds.

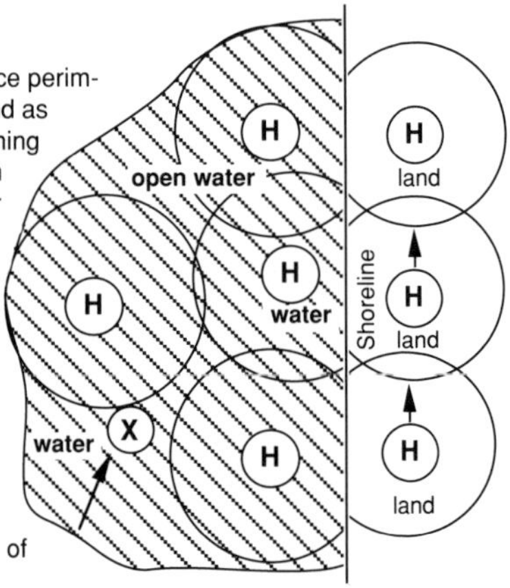

Figure 10. Water beams versus land beams (shoreline area)

beam influence subtly dominates the course of directional guidance. (See Figure 10.)

- The shoreline (water) beam investigation revealed yet another dimension of its nature. This directional beam is capable of affecting a beamer's detection of the presence of water in other than shoreline areas of open water bodies. In beaming test procedures conducted in shoreline tests previously described, the following was found:
  - Pipes or other containers, carrying or holding water, induced the same biobeaming reflex actions as described in prior shoreline (water) beam tests.
  - These water-bearing containers can be on or beneath the ground surface.
- When over water, but beyond a 45-degree angle plane from the water's edge (shoreline), the over-water beam influence on a beamer's reflex is 0% BRI in all directions faced:
  - During the nonseasonal migration beam periods
  - When no compatible beamer's color-value over-water homing beam source has been established within effective range.[7]

[7] The on-course reflex actions of a beamer when approaching vertically to water-filled pipes or hoses induce the same previously described reactions as when a beamer conducts tests at the edge (shoreline) of any body of water. Such beaming tests conducted near water pipes or hoses led to the following conclusions:

- Once a beamer's over-water homing beam source has been established, and the beamer is within an effective radius of this source, a change occurs that now parallels the over-land directional beam patterns.
  - Now a beamer's detected reflex is one of 0% BRI when facing the established homing beam source.
  - The beamer's detected reflex is 100% BRI when facing in all other lateral directions.
- Specific water-beaming tests, which are on record, indicate that when a homing beam source is established over a body of water, it is the floor of the water area upon which the homing beam source is established. Based upon the results of such repeated beaming tests, the term *shoreline beam* (or water beam) becomes self-descriptive. No attempt is made to explain such phenomena. The sole purpose is to recognize it as an important facet in the complex study of lateral directional biobeaming.[8]

---

[7] continued:

1. Seasonal variances exist in shoreline beam orientation patterns.
2. Shoreline beams have lateral orientations (northward or southward) in relation to the lateral orientation of an existing seasonal migration beam.
3. Bidirectional shoreline beam orientations exist during the quiescent migration beam seasons.

[8] All over-water biobeaming data is the result of repeated beaming test procedures, conducted either

1. When standing in a small flat-bottomed boat while in open water
2. While standing on land, above a shallow underground water container, the diameter of which is ample to allow a beamer to stand beyond a 45-degree angle plane from the water's edge.

Beamer's directional antenna influence factor

In the multifaceted phenomena of biobeaming study, no problem has demanded more time and effort than the beaming procedure referred to as the *beamer-to-ground contact factor* (i.e., beamer's stance procedure). Although the procedure has since been abandoned, this factor has been recognized as an influence on directional beaming detection since early investigations of biobeaming phenomena. Evidence can be found in work notes from the past twenty years and more recent records in the form of daily logged beaming results. The often used terms *beamer's terra firma stance* and *beamer's cushioned stance* in past records reflect the years spent in studying the unstable differential variances in diurnal BRI readings resulting from beamers' contact-to-ground beaming procedures.

Constant probing for better understanding of such beam disruption patterns eventually led to a more recent discovery that has completely eliminated the erratic directional beaming results reported in all earlier contact-to-ground experiments. This important finding established beyond any doubt the existence and sensitivity of the biobeaming phenomenon *now* referred to as the *beamer's bioantenna influence.*

What does this term mean, and was ignorance of its existence responsible for years of biobeaming tests resulting in varying, unstable beaming test data?

It is important to answer these questions so that other curious individuals, intent on becoming involved in biobeaming procedure, can experience uninterrupted, accurate beaming test results. The following can be observed during normal beaming procedures using the directional beaming indicator. (See Chapter 4.)

- A beamer's bioantenna geomagnetic beam reception is a critical factor in overall directional beam study, involving the beamer's bineural system. (See Chapter 6.)
- Specific beaming tests repeatedly indicate that the geomagnetic beams are received into the bineural system through a form of *divided (di-pole) biological antenna.*
- This divided antenna is normally represented by binocular reception in unison with any biological both left and right extremities used in beaming procedures.[9]

Discovery of this basic comprehension regarding the beamer's bioantenna phenomenon is a critical breakthrough. The improved directional biobeaming procedure rewards the beamer with consistent detection of existing stable directional beam orientations.

Eliminating the intermittent beam disruption factor through better understanding of the directional antenna influence factor results in the following:

- Complex contact-to-ground stance requirements are no longer necessary.
- Any form of beamer's bipodial stance during biobeaming test procedure will result in a beamer's maximum 100% guidance ability throughout both

[9] Such extremities should not be entirely insulated from *light* by clothing, etc.

diurnal and nocturnal directional beaming periods.[10] (See Chapter 7.)

- Diurnal biobeaming detection, now found to be a constant factor, greatly expedites all facets of beaming test procedures and study, due to the new consistencies in biobeaming test results.
- When any electrically conductive metal is exposed and in direct contact with both left and right extremities or extremity, the same 100% effective beam detection will result (i.e., 100% BRI facing off-course and 0% BRI when facing on-course direction).
- If, however, the metal contacts only a beamer's left *or* right extremity or extremities, unstable, erratic directional beaming will result. This condition is referred to as a *beamer's bineural antennae imbalance.*[11] In short, metal accessories (for example, a wrist watch, ring, etc. that acts as an antenna) on a beamer are bineurally *balanced* when the metal is in direct contact and exposed on both left and right

---

[10] Elimination of the intermittent beaming disruptions, however, will not cancel the directional beaming disrupting effect of specific conditions existing in close proximity to biobeaming procedures in progress. (See the section entitled "Other Beaming Disruption Forces," in this chapter.)

[11] Human beamer test results consistently indicate that the metal tagging or banding of migrating/homing birds or other animals may well be affecting their directional beam detection—unless such tagging or banding is applied to both left and right extremities (e.g., legs or wings, etc. of the beamer).

extremities. A beamer also may not wear exposed metal objects in direct contact with the left and/or right extremities.[12] In all circumstances, however, such on-course detection is possible only when a beamer directly faces the direction of existing beam influence.

A beamer's bioantennae influence is not affected solely by electrical conductive metal objects. The specific facing direction of a beamer's head and focal alignment also are critical beaming factors. In order to affect this critical facet of directional beam detection, a beamer's head and bineural focal direction must be aligned head-on with the direction of that existing directional beam orientation. In this case, a beamer's bineural antenna reflects the interaction between binocular focus and the area near the beamer's left and right eyebrow area.[13]

[12] The subtle cause for experiencing years of erratic biobeaming test results points to the following facts:

- When my initial biobeaming discovery was made, I was wearing metal accessories consisting of a wrist watch and ring.
- Unfortunately, due to habit, both my watch and ring were worn on the left wrist and hand, which unknowingly was responsible for the erratic directional beaming results that subsequently followed, as a result of that referred to as beamer's bineural antennae imbalance.
- Had the watch been worn on the left wrist, for example, and the ring (or any metal accessory) on any right side extremity, or had no such adornments been worn at all, normal (100% efficient) biobeaming results would have been detected.

[13] Not by happenstance was the revelation that a beamer's bineural directional antenna involves head-on alignment of both beamer's body and the beaming indicator (binocular focal/eyebrow factor). The realization evolved from years of countless tests that precipitated this discovery.

In conjunction with the bioantennae characteristics of a beamer's left and right extremities, a beamer's retinal area as bineural light wave receptors is yet another integral part of the overall bioantenna study. This is demonstrated during biobeaming results under the following specific conditions:

- With all light blocked from only one or both beamer retinas, total loss of on-course detection (0% BRI in all directions faced) will result. This repeated loss of detected beaming reaction clearly establishes that the beamer retinas, as ocular antennae units, play an important role in the overall role as receptors of directional beam influences.
- Figure 5 shows that a beamer's detected migration beam BRI will be partially disrupted when a compatible color-value established homing beam source influence lies within the *peripheral* range of a beamer of like color-value status.

The retinal antennae, however, are effective as true directional antennae only when operating in unison with the beaming indicator. The indicator is a simply designed tool that serves not only as a directional antenna, but also as an indicator of the degree of BRI variations during biobeaming procedure tests.

Diverse factors have been found to act as effective beamer directional antenna. These factors are as follows:

- Binocular vision
- Dual eyebrows
- Dual body extremities
- Dual worn metal accessories contacting body and exposed to light

All factors require a balanced left/right beamer characteristic. This reinforces the theory that all directional beaming antennae design reflect di-pole characteristics.

4 Directional Beaming Indicator

Indicator design

The basic beaming procedure presented in Chapter 2 briefly outlines the beaming indicator's purpose and use. This chapter further explains the purpose of the beaming indicator's design.

As previously stated, an indicator is used in all directional beaming procedure, but it is *not* necessary for directional beam detection. Its use, however, amplifies beam detection efficiency by increasing both the sensed and visual degree of registered beaming reflex action induced during beaming operations. Figures 1a and 1b (see pages 10 and 11) show that the indicator can be held at the ends between one finger of each hand by the inward pressure while the arms remain relaxed at the shoulder joints for totally relaxed pendular *cocking* arm motion. The indicator is held while a beamer's elbows pointing outward are bent at approximately 90 degrees. This inward holding pressure on the indicator is the result of gravity pulling the bent arms downward and therefore inward (tong fashion) from the shoulder joints. It is important to hold the indicator with end contact points at about waist level due to the 90-degree angle of the arms at the elbows. This level is critical because only at this point can a maximum degree of BRI reflex be detected.

Holding the indicator using this procedure effects two conditions:

1. Contact between the hands brings together the circuit of the beamer's left and right bineural system. (See Chapter 6.)

2. The inward pressure on the ends of the indicator produces a *collapse tendency* or stress within this curved instrument.

These factors are critical to biobeaming reflex actions, showing a relation to the stress condition that is reflected in the indicator's curved or vee design. This design renders a more pronounced collapse tendency from applied inward force at the indicator's ends.

Figure 11a shows the indicator properly held with arc or vee pointing downward. The gravity forces and stress-collapse tendencies are in harmony of direction, thereby encouraging stress-collapse tendencies. Figure 11b shows gravity forces and stress-collapse tendencies in opposition, thereby diminishing stress-collapse tendencies. No directional beam reflex action is possible when the indicator is held with the arc upward. This strongly suggests that gravity forces acting upon the curve is among the influencing factors involved in the beaming indicator's effective operation. Since the downward position is the only direction in which an indicator can be held to induce beaming reflex action, the premise is as follows:

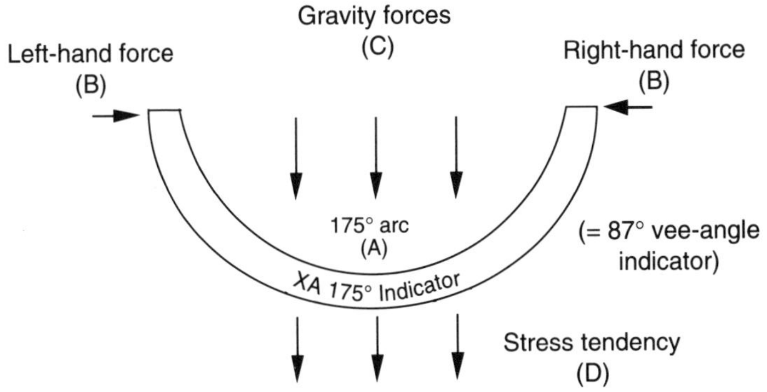

a. Proper beaming position showing gravity forces and stress-collapse tendencies in unison

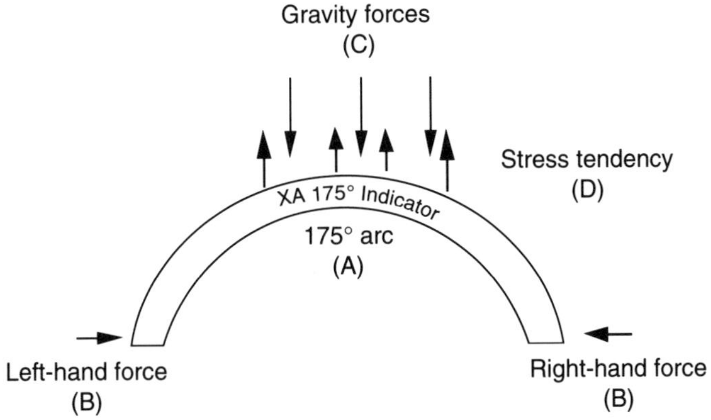

b. Improper beaming position showing gravity forces and stress-collapse tendencies in opposition

A = 175° arc design used with all beams
B = Inward force from hand-held indicator
C = Gravity force
D = Stress-collapse tendency from (B) inward pressure

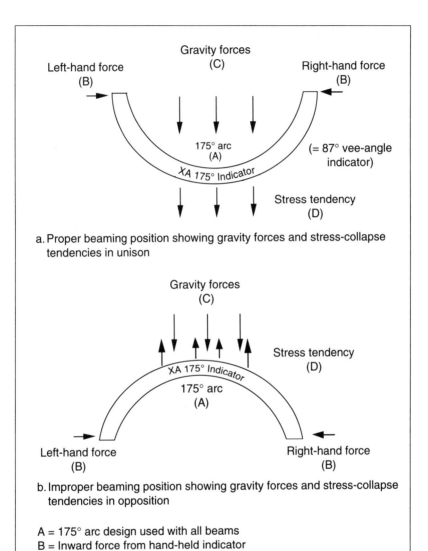

Figure 11. Beaming indicators.

1. Indicator stress force direction, in concert with gravity, effects unrestricted collapse tendency in the indicator. It is apparently necessary in inducing directional biobeaming reaction.
2. Figure 11b shows that the stress force in the inverted indicator is opposed by gravity forces, restricting total collapse tendency. Beaming reflex actions are then undetectable. Under such conditions, the beamer's reflex action falls to 0% BRI in all directions faced.

Figure 12 shows the indicator design, which is curved with approximately 175 degrees of arc. The illustrated indicator angle has proved to be efficient through years of daily hourly diurnal beaming tests. The indicator is calibrated in increments of 25% to indicate the degree of percent of BRI (beamer's reflex intensity) ranging from 0% to 100%. These markings can point out any detected differential reading fluctuations in BRI. This reading represents the beamer's guidance factor during homing and migration beam tests of on-course versus off-course BRI.

Further discussion of the beaming indicator design will detail the importance of design in arc degrees in relation to a given indicator's efficiency.

During an uninterrupted directional beaming period, any indicator design that fails to induce in excess of 50% BRI, with the beamer facing toward an off-course direction, and less than 50% BRI when facing an on-course direction, is an indicator of no beaming value. This is due to the need for detection of a BRI differential. Tests employing a variable pitch (vee) indicator show

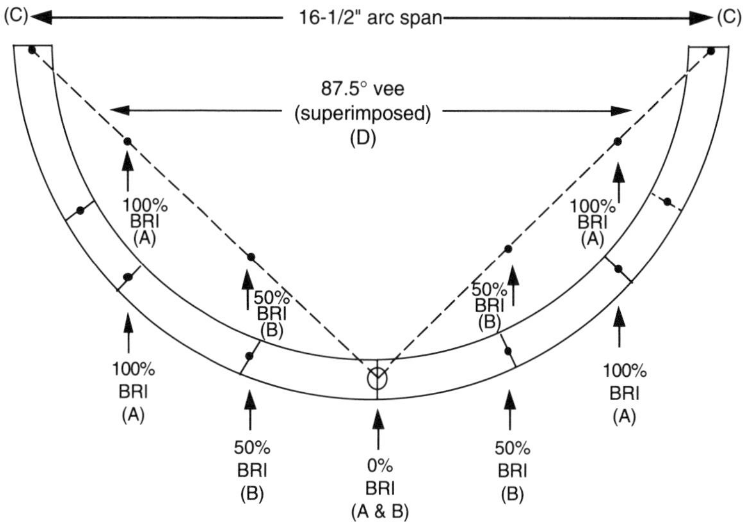

A and B (left or right) = Calibrations indicating two variations in the degree of "off-center locking" of indicator through bioreflex action (recorded in BRI percent)

C = Exemplary XA 175° indicator span

D = 87.5° vee angle indicator design (dotted line) superimposed on XA 175° indicator illustration

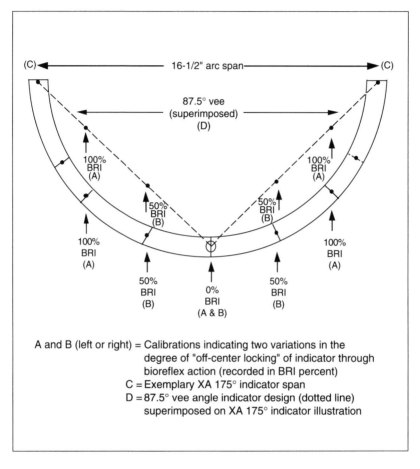

Figure 12. XA 175° vee indicator versus 87.5° vee indicator.

that the degree of indicator effectiveness is relative to the degrees of the indicator's design arc or angle (Figure 12). The following results were obtained when using indicators of different spans:

- A curved indicator of approximately 175 degrees of arc is the most effective directional beam indicator design. This specific degree of arc accounts for the strongest 100% off-course BRI reflex action in a beamer. It therefore requires the most widely spaced calibrations between 100% BRI and 0% BRI.

- Using an indicator design of between 175 degrees and 116 degrees of arc, the decreased efficiency of reflex action between 100% BRI and 0% BRI begins to narrow relative to the degrees of reduced arc in the indicator design.
- An indicator design of approximately 116 degrees of arc induces in the beamer the same BRI of 50% in all directions faced.[1] Therefore, no directional beaming is possible since on-course versus off-course 0% BRI differential do not exist.
- From 116 degrees of indicator arc to approximately 58 degrees of arc, the no-differential BRI beamer's induced reflex intensity weakens in relation to the reduction in the indicator's designed degree of arc. This BRI loss reaches 0% BRI in all directions faced at approximately 58 degrees of arc.
- Below 58 degrees of arc, BRI remains at 0% in all directions faced.
- Any arc in excess of 175 degrees (approximately) also induces a BRI reflex of 50% in all directions

[1] When referring to *no differential* or *no BRI differential* indicator readings, such readings are considered the same in all directions faced by the broadside of the beaming indicator (also referred to as *faced by beamer*).

faced by the beamer. With no BRI differential, directional beaming detection is impossible.

- Any workable beaming indicator also serves as a directional antenna, working interdependently with the directional retinal antennae.

Vee-shaped directional beaming indicators (see Figure 12) are equally efficient tools. I prefer the arc indicator for testing. It is, however, a matter of choice. The fact that either design serves the same purpose broadens the field of potential animal skeletal arrangements to act as bioindicators in homing and migrating animals.

The span of an indicator from end to end determines the measured degree of off-side locked-in motion between 0% and locked-in 100% BRI. Although this means an indicator of greater span registers more discernible action, the most comfortable span for a beamer to handle is still the most efficient. I recommend a span of approximately 22 inches.

Figure 13 better illustrates the varied effectiveness of the resulting BRI in relation to variations in arc in the curved indicator. Note that vee angle A-B is approximately 87 degrees (100% beaming efficiency, or BRI), which is equal to 174 degrees in the curved indicator. Vee angle C-D is approximately 29 degrees (0% beaming efficiency, or BRI), which is equal to 58 degrees in the curved indicator. Vee angle A-B minus C-D equals approximately 58 degrees. The beaming range is between 100% (maximum) and 0% beaming efficiency, which is equal to 116 degrees in the curved indicator. The differential in BRI detection between A-B (100%) and C-D (0%) is relative to the differential in vee indicator angle between these two points.

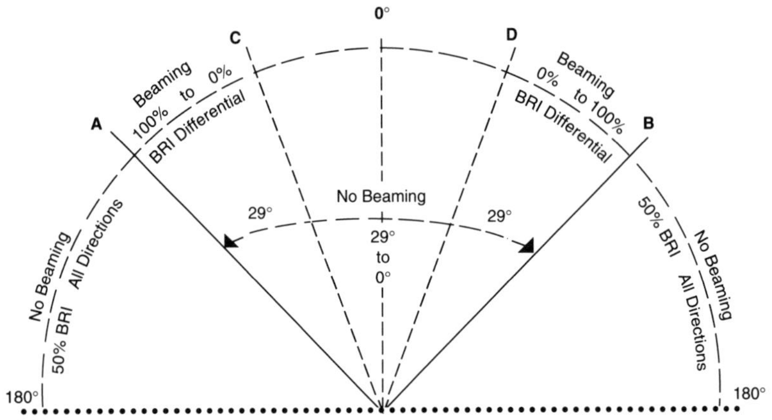

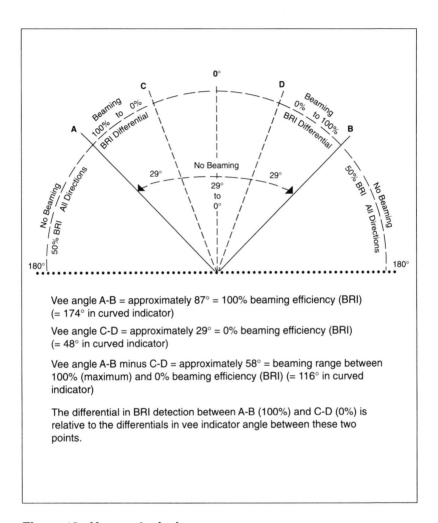

Vee angle A-B = approximately 87° = 100% beaming efficiency (BRI) (= 174° in curved indicator)

Vee angle C-D = approximately 29° = 0% beaming efficiency (BRI) (= 48° in curved indicator)

Vee angle A-B minus C-D = approximately 58° = beaming range between 100% (maximum) and 0% beaming efficiency (BRI) (= 116° in curved indicator)

The differential in BRI detection between A-B (100%) and C-D (0%) is relative to the differentials in vee indicator angle between these two points.

Figure 13. Vee angle design.

Beamer reflex guidance factor and its relation to animal homing and migration

As shown in Figure 1a, a beamer's on-course versus off-course reflex intensity (BRI) differential is the guiding factor where directional beaming is involved. Through biobeaming influence, a migrating or homing animal could involuntarily respond to such beam-induced reflex action. In so doing, the beamer would be provided with directional beam guidance through an induced sense of security when traveling in one specific direction (i.e., the on-course direction). This reflex action is referred to as *lateral directional equilibrium* and analogously parallels the unerring course of any animal traversing a theoretical concave trough on foot in total darkness. Vertical stance disturbances, in this case, would constantly correct any off-line drift.

Since the hand-held indicator is instrumental in directional beam detection, one may wonder how the detected directional beams could relate to animal homing and migration patterns. Considering the simplicity in indicator design, it is not difficult to theorize that animals, in general, carry such built-in biological "instruments."

Any portion of an animal's skeletal arrangement that is curved or angled in a downward traveling position could potentially serve as a beaming indicator, providing the degree of curve is within proper range. One specific bone structure would, of course, be assumed to serve this function.

The bird, for example, carries a most obvious and well-designed indicator, referred to anatomically as its clavicle or *wishbone*. Repeated tests using bird (turkey) wishbones as *vee* indicators prove that this bone structure is a most efficient beaming indicator. In addition, in most animals the downward curve and position of the rib cage in travel position could well be a skeletal arrangement that serves as an effective beaming indicator design.

Humans, however, in evolving to an erect standing position, may have lost their natural beaming ability since the clavicle, for example, no longer curves downward in a traveling position but is more horizontal. No indicator tested will affect migration of homing beam reflex action in a beamer *except* when positioned with the curve or vee directed at least 45 degrees downward from a horizontal plane. For this reason, only when a person stands nearly erect and employs the use of an outside contact can the detection of directional beam forces result.

It is difficult to ignore the potential relation that a directional beam influence could have to the known patterns of animals in migration whether it be on or over land, along shorelines, over open water, or when waterborne. With the discovery of not only a seasonal migration beam but, in addition, an all-season homing beam and shoreline beam, it was realized that each could serve as an integral part of a complex directional beaming network suggesting many answers to the overall animal migration and homing phenomenon.

The effectiveness of the seasonal migration beam as a guiding factor can be compared with that of the magnetic compass in the following ways:

- It offers a general, seasonal direction of guidance when no homing beam or shoreline beam influence is present.

- The all-season homing beam influence parallels, in effect, the additional aircraft guidance offered by light or radio beacons.
- The all-season shoreline beams, in turn, could be depicted as boundary walls of limitation separating land from over-water travel. This would be accomplished through the beamer's reflex guidance deflection along the shorelines in the northward or southward direction of a prevailing seasonal migration beam. (See the section entitled "Shoreline Beam" in Chapter 3, page 42.)

Human indicator beam detection versus natural animal directional beaming

This chapter earlier explained how the human beamer is capable of detecting directional beam bioreflex action through the use of the described arc-designed beaming apparatus referred to as the *beaming indicator*. The reflex action experienced is referred to as *beamer's reflex intensity* (BRI). This discovery did not, however, relate to the possible reflex guiding effect this might have upon homing and migrating animals either airborne, on land, or waterborne.

Past years of repeated beaming tests clearly show that the design of specific skeletal structures (the clavicle and/or rib cage, for example) can well serve as biobeaming indicators. The animals, however, cannot manipulate such bioindicators as the human beamer does with a manufactured hand-held beaming indicator apparatus.

The obvious question, then, is What subtle bioreflex action would a migrating and homing animal experience to induce involuntary directional beam guidance? Added variables in subsequent beaming procedure tests gradually brought about new methods where the manufactured beaming indicator was no long necessary to the beaming procedure. A method was found that used the vee shape of the human clavicle in the same way that the arc, or vee-shaped indicator was used. With the vee design of the clavicle tilted a minimum of 45 degrees downward from horizontal, directional beam detection is possible. Such a maneuver requires the beamer to bend forward at the waist to at least a 45-degree to horizontal position.

Years of testing have established that effective beaming procedure requires the need for near horizontal traveling posture of nonhuman migrating and homing animals. This posture activates their "beaming indicator" (clavicle or rib cage), thereby inducing those subtle bioreflexes that dictate directional lateral guidance.

Precisely what are the characteristics of these subtle bioreflexes? What does the animal sense to make it so willingly respond when it so chooses? This area of research is the neurological expert's domain. To the human beamer, the reflex action is obvious—any response to an off-course influence will lock the beamer-held indicator in an off-center position, resulting in both a visibly and a physically sensed reaction. However, to the insecure migrating and homing animal the subtle reflex action of lateral equilibrium induces an apparent welcome sense of security when traveling in an on-course direction.[2] In all other directions of travel, the existing off-course electrogeomagnetic influence acts upon specific autonomic reflexes constantly, subtly repelling and therefore deflecting the insecure traveler in the direction of least bioreflex resistance (i.e., on-course). This influence, however, is not an attraction toward an existing on-course source. Instead, it is a deflection through positive BRI reflex from all lateral directions other than the existing directional on-course orientation (i.e., the line of least resistance).

This tendency is toward an existing on-course direction source under the following circumstances:

- Whenever a beamer is within a 90-degree arc of that on-course alignment. Under these circumstances, a traveler through reflex action is gradually deflected toward this on-course direction from all points within the 90-degree arc.

---

[2] Not until a recent breakthrough (August, 1989) was this report considered ready for publication. A specific human beaming procedure revealed *how* a migrating or homing animal could detect and react to directional beam influence through the benefit of its own clavicle (or rib cage) to serve as its beaming indicator. The following was found:

- Using only the tilted clavicle posture, it is possible for the *human* beamer to detect a subtle influence of the seasonal migration beam without using the hand-held beaming indicator.
- With this same procedure, all-season homing beam detection is also possible only when the homing beam is established when the tilted-clavicle procedure is followed during the establishing process. (Notes 1 and 2 on page 9).

The study of detected directional beam bioreflex variables reflects over twenty-five years of persistent testing and research. The resulting discovery showed the ability of the human *animal* to *detect* and *react* to directional beam variables(BRI), which is the premise upon which all animal migration and homing abilities are then based. The decision is no more than applied common-sense syllogistics.

- During the process of any counterclockwise veering when beyond a 90-degree arc *left* of that beam source. At this point, deflection can circle the beamer counterclockwise as much as 270 degrees back to the on-course direction. In short, when on course, the constant reflex deflection influence, due to drift, can subtly lock in any willing beamer on this course during any period when geomagnetic directional beaming forces exist.

Relation to bird homing and migration (theoretical)

There can be no other explanation for the observed exploits of homing and migrating birds (for example) than that of a *bird on a beam* of deflection. Findings that involve geomagnetic directional beaming data, as detailed in this report, will be shown to offer logical explanations for the known patterns of bird homing and migration phenomena.

Except when flying above the condition, extensive fog cover presents the most formidable hazard at shorelines for the migrators. Test results show that all directional beam influence is completely voided for a beamer enveloped in the environment of diffused light (for example, fog). A bird entering an extensive, heavy fog bank would find itself with neither visual stimuli nor directional beam guidance. Inherent instincts would possibly ground such insecure travelers for their own safety.

Nevertheless, the shoreline beam would serve as an effective all-weather protector to land birds in particular. Upon approaching open water during good visibility conditions, a flyer on a migration beam course would, through instinctive fear, veer from such a direction. With visibility below affected by fog clouds, the bird, now within effective range of shoreline beam influence, would be deflected along the shoreline and in that general direction (southward or northward), depending on the seasonal migration beam's orientation. The land bird, once on this course directed by the shoreline beam, will travel on a flyway marked with inadvertently preestablished homing beam sources of its predecessors of a like species and feather who themselves were earlier influenced along this same route in the same manner.The question arises as to how some water birds would migrate from land (where they breed), across shorelines, and continuing over large expanses of water. Water birds, with no instinctive fear of traveling over water, would have no tendency to veer from it. Once over open water, the existing migration beam or well-established homing beam sources of prior birds of the same feather (color value) would carry them on the same approximate path as those traveled by their predecessors. (See Figures 7 and 10.)

Common sense dictates that a small-brained animal, ignorant of geography, navigation, or meteorology, could in no way pinpoint a habitat of its forebears following a long and often windy journey without the aid of beam guidance in flight. A form of beaming assists in bringing aircraft to a safe landing on a fog-enshrouded landing field, the plane's highly sophisticated electronic detection system controlling the guiding mechanical *reflexes*. Likewise, geomagnetic directional beams apparently guide the bird by inducing involuntary lateral directional equilibrium through bioreflex action. Such a beam-induced condition acts upon the reflexes as involuntarily as gravity tends to keep the flyer on a horizontal keel through counterreflexes in response to a sense of vertigo.

In no instance would a beamer be enslaved to such subtle lateral beam guidance—anymore than gravitation's influence prevents willful readjustment from a given state of vertical equilibrium. The will of the migrator would permit desired abandonment of such guidance when prompted by specific visual stimuli or other causes.

As with most animals, there are leaders and followers. Presumably that's the case with migrating birds. Once separated from a mass migration, an isolated bird with inherent tendencies toward insecurity depends entirely upon its involuntary lateral guiding reflexes, riding a beam of relative security.

The relationship between the discovered patterns of biogeomagnetic beaming and the known patterns of bird homing and migration has not, to this point, emphasized the full potential of the shoreline beam bordering vast bodies of open water. This guiding beam's importance as a protective factor to the migrating land bird should not be minimized. This protection occurs in those geographical areas where the coastlines of vast water bodies and migration beams intersect. One can see the potential disaster here for an east coast/North America land bird following the influence of a southeast migration beam. Being carried out over open water during periods of poor visibility or strong winds is a great threat. This possibility, however, is minimized by the shoreline beam's protective factor, deflecting the traveler on a more southerly (autumn) course paralleling the coast, and resulting in heavy migration travel along well-marked homing beam flyways fringing both the land and water borders of that route. (See Figures 10a and 10b.) Such evolved migration paths involving the shoreline beam and both the water- and land-established homing beam sources paralleling the coast are important. They provide a broad swath of directional beam guidance in a southward fall direction, defining, for example, the well-traveled North American Atlantic flyway.

Migrating land birds approaching the west coastline of North America flying the spring, northwest migration beam are also subjected to the same danger of being carried out over open water. In this situation, their protection would be dependent upon the now northward influence of the spring shoreline beam, defining the well-established Pacific flyway of North America.

On the North American continent, the many tributaries of the Colorado River, Rio Grande, and Mississippi River, as well as those rivers and streams flowing into the Pacific and Atlantic Oceans, can be clearly depicted as areas of potential shoreline beam influence. These tributaries funnel the land migrators into the four major flyways of North America, where the guiding shoreline beams and the preestablished homing beam sources have produced well-worn migration paths leading to predestined havens of the various species. On any continent, in fact, the patterns of the principal flyways would reflect the topographical features of the general area. The combined effects of the seasonal migration beam, the all-season shoreline beam orientation, and homing beam influences would define those established patterns of migratory flight.

It would be unreasonable, of course, to conclude that the directional beams involved would protect all birds in their seasonal journeys over thousands of miles. Many land birds, in particular, would be lost mainly to storms carrying them far out to sea, but their numbers would be expendable. The physically stronger and most adept beamers would most likely survive to perpetuate the species.

Bird on a beam

Applying the three existing beaming forces described, their individual guiding influences through bioreflex action, and the theory's effect of this reflex action on the animal, the following pages pose a hypothetical journey with the migrators along known flyways of birds in general. The purpose is to reinforce realization of the potential relation between biogeomagnetic beaming discoveries and known patterns of animal homing and migration.

When relating the detected biobeaming patterns with known animal homing and migration patterns, the bird was selected as a prime example of orientation phenomena. This choice is obvious. Their feats as airborne pathfinders far exceed that of migrating marine animals. The land travelers, subjected to many geographical restrictions, classify them even less desirable as an exemplary beamer. Nevertheless, the same homing and migration beaming and conditions would serve both the waterborne and land animals, neither of which would need the protective shoreline (coastline) beam.

Young fledglings on their first migration, unaccompanied by adult birds, are found to successfully migrate thousands of miles to preestablished winter/summer havens. Their route can be the same flyway as that of their predecessors of the same species and color-value markings. On such a journey the migration beam and shoreline beam are important factors in guidance, but the homing beam stabilizes the given route. This becomes more apparent when following an exemplary bird through a hypothetical migration journey along a known flyway of a given species.

In late summer, many north equatorial land birds, for example, tend to be restless. Some species begin to gather in flocks, suggesting a growing sense of insecurity. This tendency is possibly induced by an inherent urge to abandon their summer breeding and feeding grounds. With this desire to take flight but with no sense of direction, due to the then seasonal absence of a migration beam, only the local, all-season homing beam would prevail. The homing beam would provide a guiding beam back to that summer breeding area that their approaching fall instincts now reject.

With the passing of the fall equinox, however, a change in circumstances would occur with the triggering of the now northern hemispheric seasonal migration beam to the southeast. The restless bird in its daily gyrations would now gradually adapt to this new seasonal guiding influence, finding security in a given direction of lateral flight that subtly guides it away from its currently unattractive area. Either in following flock leaders or in solar flight the northern hemisphere bird would be unknowingly guided in a southeasterly, fall direction. Its old homing beam source would no longer be an influence once the migrating traveler is oriented beyond a 90-degree arc of that direction and eventually completely beyond its effective range due to terrain or other obstructions blocking the path of the near horizontally emitted homing beam forces.

A fall migration beam with its guidance in a southeasterly direction (northern hemisphere) would, however, fail to carry the bird to the winter feeding grounds of its predecessors. Wind drift could compound this problem, carrying the traveler far from such a course. However, southeast travel (northern hemisphere) would eventually cross paths with rivers or streams—coastlines or preestablished homing beam sources of like color-value status as that of the traveler. At this point, a shoreline beam of compatible homing beam influence would intercept the bird on a general southeast course and deflect it along with its companions into those well-traveled migration flyways.

Such flyways could be followed, at times, through visual stimuli; but ornithological observations prove that such paths are often followed when the ground cannot be seen, such as above fog or flight above heavy overcast. In cases where flyways parallel waterways, visual directional flight would not be necessary. The resultant directional shoreline beams emitted upward and outward from the shorelines of these water bodies would provide fall-season reflex guidance from a point directly above such courses (with the exception of those rare instances when a shoreline would lie in an east-west direction). (See Figure 8.)

Simultaneously, with the resultant heavy traffic along the waterways, these courses would be unavoidably well marked with the preestablished homing beam sources of varied color-values established in earlier flights by bird species of their own color-value status. These homing *beacons* running parallel to shoreline beams would appear superfluous, but homing beam sources in given areas would tend to result from prior heavy traffic by birds settling in areas that indicate well-used feeding grounds and rest havens. The not-too-intelligent migrator may not, of course, interpret such areas for what they are, particularly the fledgling. However, the intervening guiding influence of any such homing beam source along with hunger, weariness, and visual stimuli, would tend to deflect the bird toward those feeding grounds as long as the homing beam source lies within 90 degrees laterally of the bird's headed position. Instances may rarely occur when no homing beam or shoreline beam influence is within range of a stray bird. (See Figure 6.) In such cases, the seasonal fall migration beam would prevail, carrying the land traveler southeast eventually to intersect with other flyways of the same species.

In some cases, however, rivers or streams course mountain canyons, offering shoreline beam guidance along with the established homing beam sources that would develop along the banks. Considering this, high mountain ranges would, in general, be more of an advantage than a deterrent by funnelling migrators into well-traveled flyways.

As with the migration beam, the shoreline beam could not alone guide all migrators on the final leg of their long journey. Many western North American birds, for example, follow the shorelines of Central America when migrating southward. Others, such as the bobolink, span the Gulf of Mexico to the Central American coast or fly to the scattered islands serving as *stepping stones* to the north shores of South America. Here the bobolink would follow the preestablished homing beam course of its forebears.

Not to be ignored, however, is the seasonal migration beam, which could also be an effective force. A bird responding to this beam could be influenced southeast, a direction that could carry the lone unwary migrator to the shores of South America. It is evident, however, that all bobolinks in migration do not travel the same identical course over the Gulf. That would be impossible, considering the unstable meteorological conditions in that area. In a strong wind, a flyer following the migration beam influence would drift with an easterly wind, without losing its southeast orientation. In this manner many drifting flyers could be lost to the storm.

It is under such circumstances that the established homing beam comes into play. These homing beam sources, unlike the migration beam, would serve as invisible lighthouses. Regardless of wind drift, such *beacons* over the Gulf unwittingly influence, more strongly, the bobolink's lateral 0% BRI guiding reflexes. The windblown traveler is deflected in that direction toward a point within range of a subsequently preestablished homing source along the southward route, or a point within sight of the islands of the West Indies or mainland of South America.

A homing beam source, over water, indicates that it was established with its source emanating from the earth below the water. (See Chapter 3.) Although the limits of its range are yet to be fully determined, its effective radius would be vast, due to the absence of high-rise obstructions. The beam could possibly terminate at that point where its emitted horizontal beam waves depart from the earth's curvature. (See Figure 6.)

Nevertheless, a land bird with a lost sense of direction over open water would react to an inherent instinct to circle. In time, through such gyrations, overlapping homing beam sources (Figure 7) would be established over the waters of the Gulf by such flyers, evolving into a network of guiding beams for later migrators of like color-value plumage to follow. These nonreasoning birds would not be aware of this feat. A bird's migratory beaming course, being involuntarily controlled through a beamer's 0% BRI reflexes toward lateral directional equilibrium, would require no more forethought than does the reflex that dictates the complexities of flight itself.

Following the March equinox, the northward spring migration of the bobolink over the Gulf waters now separating them from the south coast of North America follows approximately the same route as that of the earlier fall journey. The bobolink's original pathfinders, however, may not have found the situation any easier due to the acquired spring molt alterations in their color-value status that occurred prior to the return journey. With any possible change in feather adornment, the established homing beam sources of the prior fall migration southward would have carried a color-value status not compatible with the birds' current, if altered, plumage colors. The detection of the southward established homing beam sources is then rendered impossible. Such a dilemma, however, would not have posed any more of a problem than that on their earlier flight south just six months prior. The seasonal migration beam, now to the northwest (northern hemisphere), would have subtly steered these early pathfinders in that direction, guiding them back to the shores of North America. Storms or not, the northwest beams would not carry the land birds into open waters of the Atlantic, and many travelers would make a North America landfall.

The principal purpose in discussing this Gulf flight of the bobolink is to point out that following today's flyways of this long-distance migrator, only biobeaming's established homing beacons could eventually bring the birds to safety on land during particularly the fall flight over the Gulf of Mexico. Considering the distance of such flights across the Gulf (exceeding 1,500 miles between some points) and realizing, as well, the constant threat of wind drift over the unmarked expanse of open water, only a *bird on a beam* could accomplish such a feat.

Of no less importance is the role played at those points nearing the end of a given migration. The known flight patterns of migrating birds clearly indicate that varied species abandon general flyways. Eventually they scatter in specific directions, each following a route leading to the habitat of its predecessors where suitable food, in particular, is abundant. This display of pinpoint selective homing could be accomplished only through development of preestablished homing beam path of a color-value status compatible with that seasonal color-value status of a given species. Such is particularly indicative when it is realized that often times the fledgling migrators are the first to arrive at such points.

Specific homing beam tests strongly suggest that the efficiency of directional beaming over open water would excel that of land beaming in that homing beam sources can be established only in those areas beyond the effective perimeters of other existing homing beam sources of the same color-value wavelength. However, the overlapping range of influence of additional established homing beams of like color-value just over the next horizon could well evolve into a continuous path of guidance, the general direction reinforced by seasonal migration beam influence. The conclusion, then, is that in open water, particularly, established homing beam sources of like color-value status are widely spaced, yet ample for continuous directional guidance. (See Figures 6 and 7.)

Realizing the existence of homing beam guiding influence and then considering the phenomenal homing ability of the apparently nonmigrating, but wandering, Laysan albatross, it would be ridiculous not to relate such homing feats with the guiding reflex action induced through the influence of such homing beam sources. Repeated beaming test results indicate that land-established homing beam sources are ineffective to a beamer surrounded by water. Considering this, one might rightfully question how the land-established homing beam sources on Midway Island, for example, could influence the guidance of the albatross back to the same home breeding grounds following its yearly wandering, which carries it thousands of miles to sea from that home island.

First, the land-established beam on the island could not be the immediate guiding source. Important to consider, however, is the widespread homing beam sources established by the circling albatross on and emanating upward and outward from the ocean's floor during its seasonal foraging.

The flyway to their particular island or adjoining sandpits could have been formed, in the beginning, by those few early Midway residents who found no need to venture far from the breeding ground to forage for food during the spring and fall seasons. In so doing, nevertheless, they would accidently, through normal circling gyrations, establish homing beam sources just offshore from the island's perimeters that reflected both their spring and fall color-value plumage. After their wanderings at sea over not too extensive perimeters during the fall and winter seasons, many of the forerunners of today's flyways to Midway would, in returning to their yet visible island, leave in their wake other established homing beam sources. These homing beams would reflect the altered postfall and winter color-value status of more faded, ragged plumage, beak hues, and so forth. Over many years, evolution of such homing beam flyways would expand the perimeters of this wanderlust, expanding as well the areas of added homing beam sources both to and from land. Guidance in each direction can be thought of as a *one-way street.*

Considering the behavioral habits of animals in general, it is obvious that, whether by choice or inability, not all animals have the inherent urge to migrate. Nor do all animals reflect homing abilities. Based upon annual flight patterns, one would tend to classify the albatross as a homing wanderer, ignoring the subtle influence of the seasonal migration beams and concentrating on foraging over the expanses of open water. As the spring season nears, a natural urge beckons its return to home breeding grounds. Geographically ignorant, however, of the direction this haven may lie, the albatross, possibly in a state of restlessness, would soon find solace in reacting to the 0% BRI directional guidance dictated by its spring color-value flight reflexes. The albatross is then *homing,* guided by prior years of preestablished spring homing beam sources funnelling the wanderers back to its predestined speck of land.

Unlike the Laysan albatross, the shoreline birds, some of which are recognized as the most distant migrators, follow flyways along coastlines, abandoning the land masses at given geographical points to continue their migration over water. For example, the Arctic Tern, a shore bird, follows known routes from the northernmost regions of East Canada to the shores of Antarctica. The migration involves minimal overland travel, their main fall course heading southeast across the North Atlantic Ocean, then following western coastlines from Spain to the southern tip of Africa, and from this point southwest across the South Atlantic Ocean to areas of the Antarctic coast. Some terns, however, are known to follow the coastline around the southern tip of the African continent, onward to an area bordering the Indian Ocean.

Based upon the discovered seasonal biogeomagnetic directional beaming forces with the potential of inducing directional bioreflex guidance, a theoretical journey such as that of the Arctic Tern, for example, would cease to be classified as unexplained phenomena. It would, instead, strip the not-too-intelligent bird of prior undue credit for its navigational feats through recognition of the biogeomagnetic directional beam guiding forces found to exist.

Throughout this long migration of the Arctic Tern, the crossings from Canada over large expanses of the Atlantic would be marked with the preestablished over-water homing beam and migration beam sources of prior fall migration flights.

Once encountering land on the west coast of the European and African continents, the shoreline beam would become the principal guiding influence carrying these water birds in a seasonal southward direction along coastlines often obscured below from the migrators by solid overcast or fog. When referring to any migration flyway map that shows this area of the globe, it will be evident that such a shoreline course is well established. At the southern tip of Africa, however, the main flyway course of the Arctic Tern shifts southwest in the general direction of the Antarctic Peninsula. Why would this general southwest shift over water from the African coast occur? Why would some birds remain on that African shoreline and continue around the southern tip of that continent and northward to the shores of the Indian Ocean?

Due to the coastal orientation at Africa's Cape, a unique directional beam dilemma would exist. Prior to the passing of the December solstice, the seasonal south hemispheric migration beam would be oriented to the southwest, guiding the migrators toward the Antarctic coast. However, with the passing of the December solstice and its termination of the seasonal migration beam, late arrivals at Africa's southern tip without a migration beam for guidance would yet experience reflex detection from the then seasonal postsolstice, bidirectional shoreline beam and all-season homing beam sources preestablished from earlier tern migrations. This directional influence could guide the birds to the shores of the Indian Ocean. The tern flyway pattern of migration at this point illustrates how, in some cases, the all-season homing and shoreline beams in harmony with the seasonal migration beam influences could affect unique divergence in migrational directions.

Following the March equinox, the main known flyway of the Arctic Tern in its then northward migration would carry the Antarctic-based travelers on a preestablished color-value homing, and southern hemispheric migration beam course to the northeast. Such a route would lead to landfalls on Africa's west coast. Here, the shoreline beam, now in an influencing seasonal northward direction, would hold the birds on a shoreline course to a point on the coast of Spain where the shoreline veers to the east. Here, again, shoreline beam detection would be minimal due to its east-west orientation. At this point, the tern's spring flyway veers over the Atlantic Ocean on its now north equatorial northwest seasonal migration beam and within the influence of preestablished homing beam sources from prior spring flights. Such a course would carry the weary travelers, yet unwittingly, back to their predestined summer breeding grounds of North America.

With the discovery of the seasonal migration beam—as well as the all-season shoreline and homing beams—it is apparent that these individual directional beams cannot be overlooked for their potential in inducing biobeaming guidance in patterns paralleling those known homing and migration patterns of the animals, whether in the air or on water or land.

5 Interaction of Biogeomagnetic Forces

The LMR apparatus (light, magnet, rotation principles applied)

Early investigations showed how biobeaming reactions would be negatively affected by a bar magnet that is not oriented with Earth's magnetic field and placed near a beaming test area. (See Figure 3.) This basic discovery evolved through years of experimentation into the design and construction of an uncomplicated test apparatus that would possibly simulate Earth's seasonal directional migration beam forces. It was reasoned that such a test apparatus would eventually aid in identifying the electromagnetic/geomagnetic forces strongly suspected to produce the geomagnetic beams detected and under study.

The original apparatus (since revised) was designed to permit horizontal light to act upon the separate magnetic poles of a bar magnet placed in natural orientation with Earth's magnetic field. This was accomplished by placing the magnet horizontally within a cylindrical, nonmagnetic housing. The poles of the magnet were exposed to horizontal light by means of 1/4" apertures located so that light was directed toward the bar magnet's poles. A method (later revised) of controlling the light exposure to either one or the other of the poles was accomplished by using a rotatable nonmagnetic shield that rested in a pivot point atop the LMR housing.

In operating LMR test apparatus, the normal seasonal geomagnetic beams were successfully duplicated. However, certain instabilities developed under repeated testing that indicated an apparent error in the design of the apparatus.

This apparatus gradually evolved into a design now successfully used in varied directional beaming simulation tests. The current apparatus is referred to as the *LMR* (light, magnet, rotation) beam simulator. In Figure 14, a bar magnet rests vertically at the exact center of the LMR housing so that it is aligned at the center of an axial line directly below the rotor pivot point. The black/white cylindrical reflectors stabilize the light intensity source to the magnet's poles. The insulation of the LMR stand is the direct ground contact. The theory upon which the design is based, as well as its operation, is explained in the following pages.

The LMR test apparatus, as a simulator of certain geomagnetic beam forces, embodies several important purposes. Its capabilities in simulating at any time the normal seasonal migration beam forces have rendered this apparatus indispensable. For example, prior to LMR usage, one full year was required to study one complete cycle of the Earth's seasonal, directional migration beam patterns. With the apparatus, an entire cycle can be duplicated within a few minutes. With tests in the northern hemisphere and with the beamer standing north of the LMR's *equator line*, the mere rotating of the LMR rotor in a given lateral direction, counterclockwise or clockwise, will produce a detected direc-

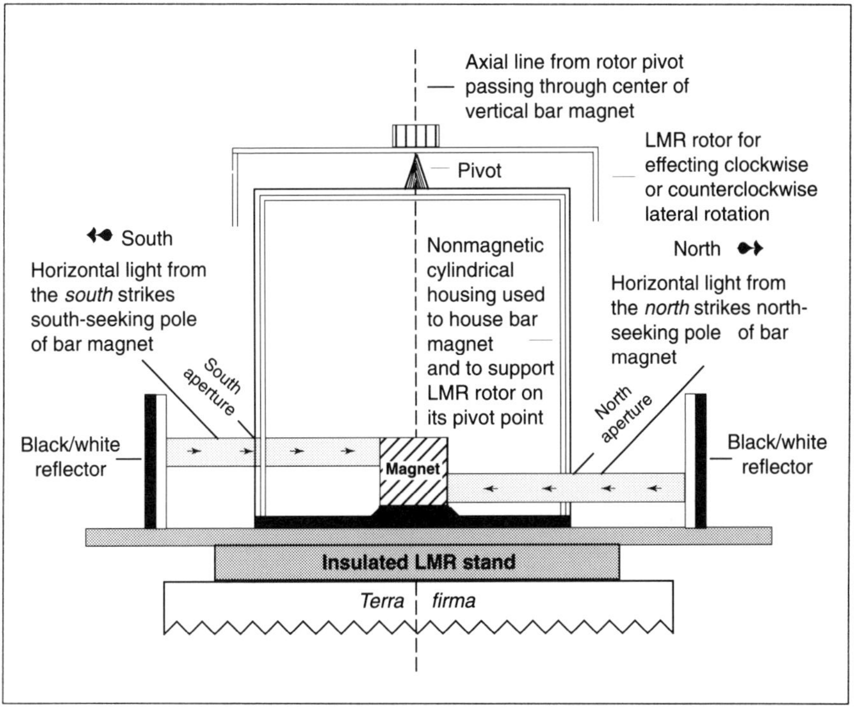

Figure 14. LMR Apparatus (cutaway view).

tional beam either to the northwest or southeast. Such controlled beam orientation is relative to the lateral direction of rotor rotation applied.

The LMR apparatus duplicates normal, seasonal geomagnetic forces. By interacting with certain bioreflexes, this apparatus can trigger an induced biobeaming reflex action of the beamer involved. To accomplish this LMR test procedure, other factors must be included that are integral parts of the overall LMR operation.

This chapter contains the findings and theories that have resulted from extensive tests involving LMR apparatus study. These tests and results explain the purpose of the LMR design factors that enable this apparatus to duplicate the normal seasonal migration beam forces. These forces can be detected within a 100-yard radius of the LMR beam simulator.[1]

The intention is to demonstrate that in LMR design the principal factors applied (i.e., the LMR's magnetic field, light interaction, and relativity of LMR's rotor rotational direction) give valuable substance to the theory that these same interacting factors trigger Earth's seasonal migration beams, identified in this book as a facet of the geomagnetic beaming phenomena.

LMR design and operation

The currently designed LMR apparatus (Figure 14) encloses a one-inch bar magnet in a cylindrical nonmagnetic housing that stands approximately eight inches high and four inches in diameter. A rotor rests on a center pivot point on top of the housing. It is skirted for no other purpose than to stabilize rotor balance. Both rotor and housing are painted black on both inner and outer surfaces.[2] Small opposing light apertures (1/4" in diameter) in LMR housing are staggered to allow light to enter and act upon a

[1] Considering the limited radius of a small bar magnet's ambient field, the fact that the LMR migration beam simulator extends its radius of detected influence to 100 yards is phenomenal. It is theorized, however, that this effective distance of influence is not directly that of the bar magnet's ambient field, but a 100-yard radius of directional beam influence resulting from interaction of light acting upon that magnet's magnetic field. In so doing, it duplicates Earth's normal directional beam forces as detected during normal beaming procedure.

[2] The black shading of the LMR apparatus stems from earlier discoveries. It was found that any loose magnets near beaming tests can disrupt normal beaming test results. Only when stored in all black nonmetallic containers will such magnets be nondisruptive.

bar magnet's magnetic field in the area of its north- and south-seeking poles. This bar magnet rests inside the LMR housing and is positioned vertically, yet it allows normal geomagnetic polar orientation at all times with Earth's magnetic field.

To further describe the operation of LMR, the basis upon which it was designed must be explained. The following two major factors were theorized to interact to produce Earth's seasonal directional migration beam influence:

- Earth's solar light/magnetic field interactions
- Earth's apparent rotations direction as *seen* from the sun[3]

That these factors are involved in geomagnetic directional beaming is patently reinforced by the ability of LMR's design to repeatedly duplicate Earth's seasonal directional migration beam influence when properly applied.

Light/magnetic field interactions

Since early experiments with the LMR beam simulator, a natural reaction was an effort to understand the specific action of the individual forces suspected to trigger directional beam influence. The first approach was to attempt to find a correlation between the intersecting light angle of attack with the magnetic field of LMR's bar magnet and the angle of solar light intersection with the Earth's polar magnetic field.

---

[3] LMR lateral rotational direction is a necessary procedure in triggering a given seasonal migration beam orientation.

This approach was based upon the characteristics of the *Zeeman-Faraday effect*[4] since the polarized light beams resulting from the interaction of light passing at right angles through a magnetic field suggests a parallel with the sun's everchanging angle of attack upon the Earth's polar magnetic field. This condition was suspected to possibly trigger directional beams of a polarized nature, which are detected through biogeomagnetic directional beaming procedure.

The subsequent discovery that beam detection ability is totally voided (0% BRI) when a beamer wears Polaroid lenses adds substance to the theory that polarized light is a basic factor in directional beaming phenomena study.

To investigate the light/magnetic field interactions involved in LMR design, careful experiments were conducted in an attempt to pinpoint the specific area where light, passing through the apertures of the apparatus, intersected with the vertical bar magnet's ambient field. This experiment was approached by masking the LMR housing apertures, thereby reducing the openings to ap-

---

[4] The term *Zeeman-Faraday effect* stems from the original discovery by Michael Faraday (1791–1867), which was later expanded upon by Pieter Zeeman (1865–1943). It was Faraday's finding that a magnetic field can affect the polarization of a beam of light. Light consists of electric and magnetic fields moving through space in the form of waves. If the electric field always lies in the same plane, for example, the direction of propagation is perpendicular to it and the magnetic field is perpendicular to both. Light of this nature would be emitted by an electric charge oscillating up and down and is said to be *plane polarized.* Faraday later found that in a variation of light/magnetic field tests, such light polarization assumed a different form, referred to as *circularly polarized light.*

Zeeman, in later research, added new dimensions to this finding with the discovery that not only the propagation but the emission of light, as well, is affected by a magnetic field. (Zeeman effect, *Collier's Encyclopedia.*)

proximately 1/8" in diameter. These smaller apertures could be moved both vertically and horizontally. In so doing, it was then possible to direct horizontal light to more specific areas of the bar magnet, ranging between its equator and the pole face areas. The following was found:

- Light/magnetic field interactions involved in LMR apparatus design are produced regardless of the extent of a given bar magnet pole area exposed to horizontal light.
- Only when light is purposely diffused at LMR apertures is the light/magnetic field interaction disrupted, negating all beamer's directional beam detection (0% BRI in all directions faced).

The finding that light, when interacting with the magnetic field of a bar magnet, can simulate Earth's geomagnetic direction beams under specific procedure is, however, only one factor upon which the LMR beam simulator is designed. The following pages expand on this factor with discussion of the Earth's *polar blink* and rotation factors reflected in LMR design and operation.

The georotational polar blink factor

The discussion of light and magnetic fields, to this point, has referred only to light of unvarying intensity and its interaction with a bar magnet's magnetic field. It was found, in addition, that another effect of these interacting forces occurs when light of given approximate intensities does not act upon both poles simultaneously. This exposure of light intensities alternating between the magnetic poles is identified here as the *georotational polar blinkfactor,* an action constantly affected by Earth's rotation from west to east.

Discovery of the polar blink factor and its influence in activating the LMR migration beam is most relevant, since the LMR test procedures and results reflect the pattern of migration beam changes detected during normal seasonal geomagnetic biobeaming periods. As stated in earlier notes, the seasons between the solstices and equinoxes are periods when no geomagnetic seasonal migration beam detection is possible. These quiescent, no-migration beam periods are primarily related to the everchanging seasonal polar light conditions that result from a combination of the Earth's rotation and the sun's changing seasonal ecliptic positions. This concept is based on the following facts:

- The ecliptic latitudes at the near approach of the solstices mark the termination period of the preceding three months of seasonal migration beam influence. Simultaneously, the solstices mark termination of that period when Earth's rotation has been responsible for a daily sequential blinking of the north and south magnetic poles.
- In place of the daily polar blink, the ecliptic position, in reaching a given degree of north latitude, exposes the north magnetic pole to constant, circumpolar solar light. Simultaneously, at the south magnetic pole, its daily polar blink is also temporarily terminated and enveloped in constant

semidarkness. Such an occurrence involves a change in Earth's normal blink pattern as it exists during those periods other than at the solstices. The normal alternating direct solar light exposure of the magnetic pole areas is then thrown out of sequence due to the summer/winter ecliptic positions.

Solstice-induced disruption of the alternating solar light exposure in the area of Earth's magnetic poles is affected by the differential in longitudinal location of these poles and the west-to-east direction of Earth's rotation in the following ways:

- Except in the solstice periods, direct solar light strikes the north magnetic pole area before reaching that of the south magnetic pole. Consequently, the sun also sets on the north magnetic pole prior to that of the south magnetic pole. Such action affects an alternating exposure of light/darkness to these areas. The normal georotational sequence in a 24-hour period beginning, for example, with sunrise at the north magnetic pole, is as follows:

	(1) morning	(2) noon	(3) evening	(4) night
North pole	light	light	dark*	dark*
South pole	dark*	light	light	dark*

*Reference to *dark* as a polar area light condition is used as a relative term (indicating less light intensity than that of the opposing magnetic pole area in a given time frame).

Interruption of this polar blink sequence during solstice periods, terminates the prior three-month period of migration beam detection. Following the solstices, the ecliptic, in reversing its course southward, soon reactivates the normal polar blink sequence.

- Migration beam detection, however, fails to recur until the passing of the next equinox three months later, at which time the postequinox migration beam is then detected to be oriented in a reversed (180-degree) direction. This phenomenon is theorized to result from the solar ecliptic passing into the opposing hemisphere. In so doing, it reverses the apparent direction of Earth rotation as witnessed from the new solar position.

To reinforce these geomagnetic beaming observations, tests involving LMR apparatus repeatedly show that *with such light exposure sequence alternating between the bar magnet's poles, a parallel exists between LMR test results and normal geomagnetic beaming patterns*.

Tests with LMR apparatus clearly illustrate that with the light exposure sequence alternating between the bar magnet's poles, undisrupted LMR-produced migration beam detection prevails:

- Providing LMR rotor rotation is applied in the *same* direction following each subsequent manual light blink change[5]

[5] In testing with LMR apparatus, it is not necessary to execute this light-blink sequence with subsequent rotor rotations as long as the sequence is not altered or advanced.

- Providing no alterations in this normal polar light blink sequence occur

Of equal importance, however, is the fact that at any point during a light blink sequence cycle, when the direction of LMR rotor rotation is reversed, the resulting migration beam orientation is also reversed.

A parallel between the following exists:

1. The effects of postequinox apparent georotational reversal *as seen from the sun*
2. The like effect resulting from reversing LMR's rotor rotation direction.

Both of the above trigger a 180-degree reversal in migration beam orientation in a given hemisphere.

In the event that LMR's normal light blink sequence is disrupted during LMR tests, only simultaneous exposures of darkness to both magnetic poles, followed by exposures of light-dark or dark-light sequences, will reinstate normal LMR-produced directional beaming conditions. Such procedure reflects the normal georotational polar blink conditions described earlier in this chapter.

The following pages clarify the alternating of presolstice polar blinks produced by Earth's rotation and the solar ecliptic to a greater extent by describing Earth's polar magnetic blink sequence during a one-year seasonal cycle. Figure 15 shows data compiled through extensive LMR tests in the course of simulating the Earth's magnetic pole light-blink sequence variations during the same period. Also shown are the northern hemispheric migration beam detection patterns that parallel those occurring during normal geomagnetic pole light-blink exposures.

Lines 1 through 4 represent the period between the passing of the March equinox and the presolstice period. This period illustrates the simulation of alternating polar blinks during 24 hours of Earth's spring rotation. For example:

Line 1, column b, indicates the first solar light blink (L pole). Column c, approximately eight hours later due to longitudinal location and georotation, remains unblinked (D) in relative darkness.

Line 2 illustrates the time of day when both poles are exposed to direct solar light (LL).

Line 3 represents postsunset at the north magnetic pole (D), while the south magnetic pole remains in sunlight during the same interim of Earth's rotation.

Line 4 illustrates darkness at both poles (DD). These alternating magnetic pole light exposures will continue in this sequence until arrival of the presolstice period in June. Column e shows direction of LMR rotor rotation as seen from the magnet's location below in apparatus housing, and columns f and g show the influential direction of resultant migration beam detection.

Line 5, the onset of the presolstice period, indicates that a presolstice multiple blink of a single magnetic pole (north pole, or LDL) has occurred while the opposite pole remains unblinked (DDD) during the same degrees of Earth's rotation. Such single-pole light-blink action disrupts all prior existing migration beam detection (column h) until passing of the September equinox.

Line 6 indicates arrival of the June solstice when the ecliptic terminates its northward progression, exposing the north magnetic pole to maximum constant light and the south pole to maximum relative darkness.

Period	a Line number	b Bar magnet north-seeking pole light exposure	c Bar magnet south-seeking pole light exposure	d Rotor rotation as viewed from above	e Rotor rotation relative to magnet position	f Resultant southeast beam	g Resultant northwest beam	h No beam detection	
	1	L	D	CC	C		X		Initial postequinox blink
March equinox to June presolstice	2	L	L	CC	C		X		See lines 1 through 4
	3	D	L	CC	C		X		
	4	D	D	CC	C		X		
	5	LDL	DDD	CC	C			X	See line 5
	6	LL	D	CC	C			X	See line 6
Post-June solstice to September equinox	7	L	D	CC	C			X	See lines 7 through 10
	8	L	L	CC	C			X	
	9	D	L	CC	C			X	
	10	D	D	XX	X			X	September equinox
September equinox to December presolstice period	11	L	D	(C)	CC	X			See lines 11 through 14
	12	L	L	C	CC	X			
	13	D	L	C	CC	X			
	14	D	D	C	CC	X			
	15	DDD	LDL	C	CC			X	See line 15
	16	D	L	C	CC			X	See line 16
December solstice to March equinox	17	L	D	C	CC			X	See lines 17 through 20
	18	L	L	C	CC			X	
	19	D	L	C	CC			X	
	20	D	D	C	CC			X	March equinox
	21	L	D	(CC)	C		X		See line 21

L = Light
D = Dark

C = Clockwise
CC = Counterclockwise

X = Resultant beam direction (migration beam only)
(C) or (CC) = Initial reversal on rotor rotation direction

Figure 15. LMR Test Data.

Lines 7 through 10 show the resumption of daily alternating blinks at *both* magnetic poles as the summer ecliptic moves into more southward latitudes. However, *no* migration beam resumption is detected as shown throughout the summer period (column h).

Lines 11 through 14 represent the fall season with the passing of the September equinox, an important period. Although the polar blink sequence is not altered, column f shows that a migration beam is now detected. Its direction of beamer's influence, however, is reversed to the southeast (northern hemisphere). This reinstatement of the migration beam is due to a reversal in LMR rotor rotation direction (column e) and simulates the apparent Earth's reversal of georotation as seen from the sun with its ecliptic, which is now in Earth's south equatorial hemisphere.

Line 15, a reversal of Line 5, indicates the arrival of the December presolstice period with the single pole blink now occurring at the south magnetic pole. This reversal leaves the north pole in relative darkness during the short-term blink and terminates the prior southeast, fall, migration beam detection (column h).

Line 16, a reversal of Line 6, indicates the December solstice, when the ecliptic terminates its southward progression, exposing the south magnetic pole to maximum constant light and the north pole to relative darkness.

Lines 17 through 20, a repeat of Lines 7 through 10, show the resumption of daily alternating blinks at both magnetic poles. Again, no migration beam detection exists, as shown in column h, throughout the now prevailing winter period.

Line 21 represents the passing of the March equinox, and like that occurrence during the post-September equinox period (fall), the migration beam is reinstated. Due to the reversal in the direction of rotor rotation, which simulates the reversal in Earth's rotation as seen from the sun, its ecliptic is now in the north equatorial hemisphere. With the passing of the March equinox, the reinstated migration beam direction of influence now lies to the northwest (northern hemisphere) as shown in column g. Line 2 also represents the completion of a one-year cycle in simulating Earth's yearly seasonal magnetic pole blink variations, which represents line 1 as well.

Figure 15 shows LMR-produced migration beams that parallel the pattern of the normal geomagnetic seasonal migration beams detected throughout a one-year cycle, strongly suggesting an apparent relation between the two. The LMR polar blink and rotor rotation factors reflect similar varied occurrences at Earth's magnetic poles during the four seasons of georotation, and each produce the same detected directional migration beams under the same conditions.

It should be noted that all LMR tests were conducted during normal nonmigration beam seasons. Such precaution is observed due to beam conflict that exists when LMR migration beam simulation is carried out in an environment of active normal geomagnetic directional beam influence (for example, accuracy of BRI readings is affected when conducting LMR tests for a beam in a given direction time frame when the normal beam lies 180 degrees from that orientation).

Application of rotation factor to biobeaming

The polar blink action that results from Earth's rotation is only one facet in the overall study of the rotation factor in relation to geomagnetic biobeaming. This action was first realized with the initial accidental self-establishing of a homing beam source through body rotation.

Subsequent years of biobeaming experiences and LMR test results pointed to the realization that the element of rotation is involved in several aspects of the beaming study. They are as follows:

- Georotation (earth rotation) is the catalyst responsible for the polar blink factors earlier described.
- The rotation factor is reflected in the apparent direction of Earth's rotation as seen from the sun's everchanging ecliptic seasonal position either north or south of the equator.
- The rotation factor is applied in the pivot-to-ground beaming maneuver[6] and acts as a beam-disrupting maneuver, either accidental or intentional. (See the section entitled "Pivot Factor" in Chapter 5.)

---

[6] As stated, LMR tests show that a beamer's counterclockwise rotation can establish a homing beam source or pivot action resulting both north and south of LMR's *equator line*. However, no on-the-spot tests have been made to date in Earth's Austral hemisphere. Based on LMR's dependability as a duplicator of normal geomagnetic beaming conditions, however, indications point to the beamer's rotation direction in Earth's southern hemisphere as counterclockwise as well.

The rotation factor with which the pivot maneuver is associated has undergone extensive testing throughout the years of biobeaming study. The conclusions from such investigations confirm the element of rotation as an integral part of the overall biogeomagnetic directional beaming phenomenon. No attempt is made here to explain the complex interaction of forces involved. However, rotation is a prevalent force, spanning laws of motion from the expanse of celestial mechanics to the microscopic world of the atom.

- A beamer's counterclockwise rotation procedure is necessary in establishing a homing beam source.

Based on these findings, rotation is, in fact, a critical element built into the LMR apparatus design. LMR study of rotational influence clearly reinforces the existence of a strong relation between the results of LMR tests and that of the rotation factor as it applies to normal geomagnetic beaming. Such findings reveal the following:

- LMR operation shows that the seasonal 180-degree reversal of a migration beam's direction of guiding influence can be duplicated by the lateral reversal in direction of the LMR's rotor. This relates to the apparent reversal in Earth's rotation as seen from the sun relative to its ecliptic position north or south of the equator following the equinoxes.
- In order to produce effective LMR operations, LMR rotor rotation must be aligned to keep the rotor pivot point above and in vertical alignment with the center point of the apparatus's vertical bar magnet. (See Figure 14.)
- When following the normal light-blink sequence during LMR beaming tests, any change in polar light exposure must be followed manually by rotation of LMR's rotor before any detection of possible change in beaming reflex action can be affected.
- When establishing a northern hemispheric homing beam while under LMR's influence, a beamer's

direction of body rotation is counterclockwise as it is in normal geomagnetic beaming. The same direction of rotation is necessary when the beamer stands south of LMR's equator line. (See Figure 14.)

Pivot-to-ground factor

The pivot-to-ground factor is important in biobeaming study and procedure. It cancels a preestablished homing beam source by a beamer affecting such frictional pivot action, whether accidentally or intentionally.

This procedure involves a beamer pivoting a foot, for example, that is in direct contact with *terra firma,* or any other rigid grounded mass. Pivoting demands just the slightest degree of such frictional contact (for example, partial foot twist) in a counterclockwise direction in either the equatorial hemisphere.

In recognizing its potential, such canceling action can be avoided when so chosen. There are, however, situations when canceling an established homing beam source is necessary. Making the pivot factor an advantage (for example, canceling the disruptive effect of an accidentally established, unwanted homing beam source in an area during other beaming tests).

A beamer's detected results from the following pivot procedure will further demonstrate the complexities involved in the overall biobeaming study:

1. Counterclockwise nonfrictional rotation of a beamer's body is necessary during the process of establishing a homing beam source. This same counterclockwise rotation factor, how-

ever, when frictional pivot-to-ground (pivot factor) procedure is applied on the same area, will cancel this same well-established homing beam source.

2. A northern hemispheric beamer can cancel a homing beam source only when the counterclockwise pivot procedure is executed with the beamer's left bineural side (for example, *left* foot). (See the section entitled "Bineural polarity factor" in Chapter 6.)
3. Inversely, a southern hemispheric beamer can cancel an established homing beam source[7] only when such counterclockwise pivot procedure is executed with the opposite bineural side (for example, the *right* foot). (See the section entitled "Bineural Polarity" in Chapter 6.)

Local disruptive biobeaming influences

Although the interaction of certain forces is conducive to triggering directional beaming forces, other beam disruptive conditions exist that, at close proximity, can void all detection of

[7] As previously discussed, LMR tests show that a beamer's counterclockwise rotation can establish a homing beam source or a pivot action results both north and south of LMR's equator line. However, no on-the-spot tests have been made to date in Earth's Austral hemisphere. Based on LMR's dependability as a duplicator of normal geomagnetic beaming conditions, however, indications point to the beamer's rotational direction in Earth's southern hemisphere as counterclockwise as well.

normal directional beaming reflex phenomena. These disruptive conditions can affect biobeaming results at all times and are as follows:

- A grid or any flat, woven material, such as a window or door screen, located near a beamer will disrupt a beam when facing the broadside of a beamer's held beaming indicator. Its effective range lies within approximately eight feet of the beamer. The degree of induced beamer's reflex intensity (BRI) is relative to the distance separating the two. At a distance of one foot, beaming detection is totally voided, inducing a 0% BRI (no reflex action).
- When a beamer stands within a three-foot radius of another individual, the BRI is affected. Here again, the degree of beam disruption is relative to the distance in proximity, totally voiding all BRI at a one-foot radius.
- When a beamer stands within a given range of an electrical current flow, beam disruption occurs. For example, when a beamer stands near 120 volts AC, and at right angles from the current carrying line, all BRI detection is totally voided. The effective range of this electrical disruption lies within approximately 50 feet of a 220-volt current carrying line. Within that range, the disruptive effect is at all-angle planes from that current line within a 360-degree radius.

- When a beamer stands within six feet of an AC radio tuned to either AM or FM reception, under specific beaming procedures, the following results occur:
  - Regardless of the facing direction of the beaming indicator, an AC radio tuned to AM radio reception will produce directional beaming disruption only when the beaming indicator is held horizontal to that source.
  - Regardless of the facing direction of the beaming indicator, a radio tuned to FM reception will produce the same beaming disruptions when the beaming indicator is held at any level from that source.

This radio wave influence is interesting in that both AM and FM radio signals are omnipresent in surroundings where such tests are conducted, yet only within the described maximum six-foot radius of an activated AC radio receiver do they trigger directional beam disruptions. It should be noted, however, that the AM/FM wave factor suggests a relationship in patterns to those beam reception angle planes of the migration beam as opposed to the homing beam in the following ways:

- Migration beam reception is detected to be in a given seasonal lateral orientation only when the facing direction of the beamer and broadside of the indicator are on an approximate horizontal plane. These characteristics parallel the beaming test results previously described where AM radio reception is in close proximity. (See the section entitled "Migration Beam" in Chapter 3.)

- Homing beam reception is detected to be in that lateral facing direction of a given homing beam source influence. However, in this case a beamer and indicator may both face on any plane between horizontal and near vertical, paralleling FM radio wave patterns. (See the section entitled "Homing Beam" in Chapter 3.)
- A beaming disruption occurs when any biobeaming procedure is conducted within a six-foot radius of a smoke alarm (fire warning system).
  - The degree of affected BRI detection is relative to the proximity of the test radius.
  - A total (0% BRI) disruption will result in a 0% beamer's reflex action when the held indicator is within a two-foot radius of the disrupting source.

Although these disruptive near-proximity beaming conditions may have minimal negative effect on normal animal homing and migration, data of the findings recorded in this book may contribute toward the eventual understanding of the overall biobeaming phenomenon.

6 Bineural Factor

Bineural factor defined

Detection of biogeomagnetic directional beams through bioreflex action entails what is referred to in this book as the *bineural factor*. This theory evolved from the detected bioreflex action that induces a beamer's specific autonomic reflex to react when subjected to certain existing geomagnetic directional beaming forces.

The bineural factor is described in Figure 1 in basic anatomical language. The complex nervous system in humans, for example, is controlled through the brain and spinal column. It is arranged so that certain peripheral sensory nerves of the reflex arc are isolated, dividing the sensory nerves of the left side of the body from those of the right side. This theory is evident in specific cases of brain damage resulting in the paralysis of only one side (left or right) of the individual.

Considering this theory, repeated findings through biobeaming procedure led to the adoption of the term *bineural*, reflecting the need to employ both left and right sides of a beamer in a specific physical manner during beaming procedure. This is a most important factor to ensure any beaming reactions.

In normal biobeaming procedure, one example of this need to employ both the left and right sides of the peripheral nervous system involves the use of the beaming indicator. In this case, one finger of each hand contacts each end of the beaming indicator. In so doing, the indicator closes the circuit of the bineural system. In this phase of beaming procedure, a properly held indicator will trigger within the beamer a toggling (locking) reflex action in the scapular region, impeding the hand-held indicator from falling to a relaxed centered body position. This reaction (BRI) is both sensed by the beamer and clearly visible on the indicator's calibrated surface. (See Figure 1.)

Years of additional investigation and test results further verify that the bineural factor is involved in other aspects of biobeaming action by responding as two independent entities acting upon the left and/or right sides of the body. Not to be excluded are the retinas as binocular light receptors.

In all beaming procedure, light striking a beamer's retinas is critical for all beaming detection. The following is an example of test results:

- Directional beaming detection is voided when either the left or right eye is tightly closed.
  - This blocking of light from either retina voids all directional beam influence. Any existing directional beam influence will then induce only a 50% BRI in all directions faced.

- Blocking all light from both eyes voids all beam detection by reducing a beamer's reflex action (BRI) to 0% in all directions faced.[1] (See the section entitled "Beamer's directional influence factor" in Chapter 3.)

This example of the need for binocular light reception in biobeaming leaves little or no doubt about the role played by the bineural factor. There are, nevertheless, other examples to further substantiate this bineural phenomenon.

- With a beamer's eye brows altered to show differing color value, the resultant beaming detection is the same as though one brow was entirely shielded from all light, such as 50% BRI in all directions faced, which would result in no directional beam detection.

  This example again shows the bineural aspect of biobeaming as equally affected by the color-value factor characteristics of light.

In addition to the above data involving a beamer's brow/ retinal influence upon a beamer's bineural system, the beamer's contact-to-ground (stance) factor during biobeaming procedure also reveals the influence of a beamer's bineural system:

- A beamer's bipodial contact-to-ground (stance) either directly or indirectly during biobeaming procedure can be on any surface without disrupting normal directional beaming results.

[1] Faced by beamer and broadside of beaming indicator.

- However, a beamer's monopodial stance under the same conditions will repeatedly negate all beaming detection (0% BRI in all directions faced).

Bineural polarity factor

The bineural factor, its ever-present relation to beam reflex action and those specific conditions that affect its temporary disruption, remains an unexplained phenomenon. Only through recognition and control is biogeomagnetic directional beaming possible, producing within the beamer a sense of lateral, directional equilibrium.

To further expand upon this complex bineural study, the following additional findings reveal specific biopolarity interactions between bineural contacts and existing geomagnetic forces.

In the course of investigating the interaction of geomagnetic forces within themselves, and this interaction with a beamer's bineural reflexes, yet another related phenomenon was found.

Through specific beaming test results, evidence shows that geomagnetic field polarities interact with a beamer's bineural polarities under most interesting circumstances. The following discovery was made in the course of investigating the pivot-to-ground factor. (See Chapter 5.)

- During normal northern hemisphere beaming tests:
  - Any counterclockwise pivot-to-ground maneuver actuated by a beamer's left side only. For example, left foot will result in the following:

~Permanent cancellation of any existing established homing beam source when pivot is conducted on that established homing beam spot.

~ Temporary reduction of a beamer's BRI to 0% at any other point where such a counterclockwise pivot is exercised.

- To augment this study, LMR apparatus tests separately showed the following:
  - When the beamer also stands north of LMR's bar magnet equator line, the same counterclockwise pivot procedure results parallel those described in the previous point.
  - However, when the pivot action is conducted south of the bar magnet's equator line, the same beam disruption occurs only when the right foot is pivoted in a counterclockwise direction.

The strong indications here are that a beamer's bineural system also responds to a condition of opposing polarities that exist between a beamer's left-versus-right bineural makeup and geomagnetic polarities through Earth contact.[2] This phenomenon indicates that such geomagnetic ground polarities are determined by their location either south or north of the equator.

The biopolarity pivot study is important in revealing apparent existence of a detectable polarity between Earth currents and

[2] Telluric (Earth) currents?

bineural polarities (i.e., biopolarities). This phenomenon, in addition, suggests opposing interactions between south-versus-north hemispheric polarities.

Conclusion

Vital biobeaming procedures have not been withheld from this test, since reporting of such data should serve a common purpose. New ground has now been sown. It is hoped that the potentials revealed within this biobeaming study will be recognized and acted upon by those schooled in the individual field involved, their combined efforts leading to a better understanding of a complex phenomenon.

Biogeometric beaming, as detailed in this book, for obvious reasons relates to the age-old inquiry into animal homing and migration patterns. To conclude, however, that discovery of biobeaming forces and their relation to animal homing and migration abilities implies existence of no other potentials would be an unfortunate error. Clear indications exist of an interesting link between geoelectromagnetic forces and bioelectromagnetic forces. The interaction between these two separate forces could well disclose a new field of vast potential for new questions and answers of which animal orientation through involuntary lateral reflex equilibrium is but one facet.

Appendix

Author's note

In the past years of researching Biobeaming phenomena, discoveries have surfaced other than the ones discussed in this book. Among such findings were those showing erratic beam results, which directly relate to the data discussed in the preceding text. These findings were considered secondary in importance due to earlier errors in beaming test procedures and were withheld to avoid further complicating an already complex subject. Past records of these encountered discoveries are revealed in this appendix. It is hoped that the additional data (on file) may further expedite future understanding of the biobeaming phenomenon, as well as answer other questions involving such mysteries as biological clocks, circadian rhythms, etc.

Miscellaneous related data of research value

Equatorial meridianal solar blink factor

Since October 21, 1988, a specific discovery led to a beamer's ability to detect constant twenty-four-hour directional beam forces. (See Chapter 5.)

Prior to that date, the daily records of a five-year study of directional beaming phenomena revealed a puzzling, constant pattern of daily directional beam detection loss of 50% BRI (no BRI differential in all directions faced by beamer).[1] This pattern of daily directional beam loss occurred at 8 PM (PST) and was reinstated at 5 AM (PST) the next morning, leaving a predictable daily 15 hours of potential diurnal beaming. This time span was found to extend from that period seven and one-half hours prior to and following solar noon at any meridian where beamer tests were conducted.[2]

Only in more recent years has this phenomenon, through varied investigative approaches, begun to reveal patterns of behavior leading to the conclusion that there exists not one (as earlier theorized, but two separate daily occurrences of solar light versus geomagnetic field reactions that in addition to the earlier discovery determine when direction beam forces can or cannot be detected. (See the section entitled "Georotational polar blink factor" in Chapter 5.) A second nonvariable equatorial meridianal light-blink action was subsequently found to exist.

During the remaining nine hours per deum, only a constant neutral beaming reaction of 50% could be detected in all directions faced by the beamer. The absence of BRI differential voided all possible directional beam detection.[3]

---

1 *Directional beam potential* implies that during a daily fifteen-hour time span, a beamer's ability to detect a directional beam force was always possible. This possibility, however, was interrupted by erratic periods of beam loss, due to the result of a poorly designed receiving antenna. (See Chapter 5.)

2 These tests were conducted at the Twenty-nine Palms, California, meridian.

3 These variable beaming patterns currently are still possible to detect through intentional use of earlier nonbalanced beamer bioantenna during beaming test procedure.

Such a finding prompted research into past beaming records consisting of daily and hourly logs of directional beam test results. It showed indisputable evidence that the daily fifteen-hour diurnal and nine-hour nocturnal nonbeaming periods remained within a *constant* time span. Since light acting upon a geomagnetic field is apparently a critical factor in triggering the directional beam phenomenon, how, then, did these daily occurrences remain in a constant light-time span regardless of Earth's seasonal, ever-changing solar light-time span conditions due to the ecliptic? Only at Earth's equator does the day/night solar light cycle remain within a nonvarying time span. Here hemispheric light/shadow conditions progressively creep from east to west, subjecting 180 degrees of Earth's equatorial surface to direct sunlight, and the opposing 180 degrees of equatorial hemisphere lying in a shadow of relative darkness. Equally important is the fact that *indirect* solar light can extend beyond the 180-degree curvature of the Earth, acting upon geomagnetic field meridians beyond that curvature point.

In parallel, recorded beaming test results repeatedly showed the existence of a daily, diurnal fifteen-hour directional beaming period potential. This pattern was found to coincide with the time span of the ever-advancing equatorial light blink conditions, based on the following reasoning:

- The fifteen hours of daily, diurnal detected beam potential suggested that in addition to direct solar light, the terminator's indirect light area beyond the 180-degree span would extend ample beam, producing meridians to 225 degrees.

- Simultaneously, 135 degrees (i.e., nine hours) of Earth's curvature at the equator would remain in a degree of darkness, representing the known per deum periods of no possible directional beam detection.

To date, the interaction of light and geomagnetic forces as a factor in the study of directional beaming phenomena is not understood, but it does further reinforce a relationship to findings of the Zeeman-Faraday effect. In addition to establishing a parallel between Earth's daily dark/light equatorial meridianal light blink cycles, the directional beaming time span patterns further indicate the existence of a directional beaming force triggered by the interaction of light and magnetic fields.

This account of past findings, although accurate for reference purposes, was updated on October 21, 1988, when a more efficient left/right (balanced) directional beaming bioantenna was discovered. (See the section entitled "Beamer's directional antenna influence factor" in Chapter 3.) With this subsequent revelation, the resultant stronger beam reception ability enabled directional beam detection throughout an uninterrupted 24-hour daily period.

The current elimination of the diurnal versus nocturnal beam reaction BRI differences does not negate the research values of the earlier recorded data showing erratic patterns of beaming and no-beaming periods. The instability of such directional beam detection was due to inefficient beamer's antenna design and its weak beam BRI reception. This condition, in turn, can result in weak or 0% beam detection ability, relative to the degree of poor light intensity at the equatorial meridians.

Bibliography

1. Ardrey, Robert, *The Territorial Imperative,* New York: Atheneum, 1966.

2. Matthews, G. T. V., *Bird Navigation,* second edition, Cambridge: Cambridge University Press, 1968.

3. National Aeronautics and Space Administration, *Animal Orientation and Navigation.* NASA SP262, symposium data, 1972.

   a. Lindaur, M., Martin, H., "Magnetic Effects on Dancing Bees."

   b. Keeton, William T.," "Effects of Magnets on Pigeon Homing."

   c. Wolfgang, Wiltschko, "The Inference of Magnetic Total Intensity and Inclination of Direction Preferred by Migrating European Robins."

4. Atmospheric Environment Service. Canadian Climate Center, Daily/hourly report on weather data at Arctic weather station (Alert), N.W. Territory, January and May, 1985.

5. National Climatic Data Center, Boulder, Colorado. Daily three-hour interval observations for geomagnetic activity indices, May through August, 1985.

Index

A

AC radio receiver 106
All-season homing beam 84
Alternating blinks 99
Alternating polar blinks 97
AM radio 105
AM radio reception 106
AM reception 105
Autonomic reflex 69, 109

B

Bar magnet 24, 25
Basic beaming procedure 7
Beam 16
  deflection 70
  direction 15
  geomagnetic influence 1
  intensity 15
Beamer 2, 8
  bineural antennae imbalance 53
  cushioned stance 51
  left and right bineural system 58
  reflex action 13
  reflex intensity (BRI) 2, 16
  retinal area 55
  retinas 110
  right bineural side 103
  stance procedure 51
  terra firma stance 51
Beamer-to-ground contact factor 51
Beaming disruption 106
Beaming indicator 8, 13, 55, 110
Bidirectional shoreline beam 84
Bineural 109
  factor 5, 109
  focal direction 54
  guiding reflexes 14
  polarity 104
  reflexes 112
  system 8, 52
Bineurally balanced 53
Binocular light reception 111
Binocular reception 52
Binocular vision 56
Bioantenna 52
  influence 54
Biobeaming 4, 15, 74, 79
  procedure 14
Bioelectromagnetic forces 114
Biogeomagnetic beaming vi, 2, 74
  detection 14
  directional beam vi, 37, 83, 109
  directional beaming forces 83
  lateral directional beaming 5

reflex action 13
Bioindicator 67
Biological reflex action 2
Biopolarity 104
interactions 112
Bioreflex 68
action 7, 74, 109
guidance 83
Bipodial stance 52
Bird on a beam 79
Blinking 94
Body alignment 43
Body rotation 28
BRI 2, 16
differential 13, 17
off-course 30
on-course 30
Brow/retinal influence 111

C

Circling 28
Color value 37, 111
bird homing and migration 40
factor 37, 38
status 37
Contact-to-ground 51, 52, 111
bipodial 111
Counterclockwise body rotation 18, 29

D

December solstice 84
Deflected 69, 71
Differing color-value frequencies 31
Directional antenna 63
Directional beam detection 8
Directional retinal antennae 63
Diurnal and nocturnal 33
Divided (di-pole) biological antenna 52
Dual body extremities 56
Dual eyebrows 56
Dual worn metal accessories 56

E

Earth's magnetic field 2, 24, 92
Ecliptic latitudes 94
Electromagnetic forces 2
Electromagnetic/geomagnetic forces 87
Eye brows 111

F

Faced by beamer 17
First rotation 29
FM radio 106
FM reception 105
Frequency modulation antenna 31
Frictional contact 103

G

Geoelectromagnetic forces 114
Geomagnetic beams 87
Geomagnetic influence 16
Georotation 101
Georotational polar blink factor 93

H

Hemispheric migration beam 84
Homing 84
  beacons 79
  beam 4, 21, 28, 67, 83, 106
  beam antenna 31
  beam source 103
Horizontal plane 106

I

Indicator 43
Indicator stress 60

L

Lateral directional beams 2
Lateral directional equilibrium 1, 14, 16, 65, 71, 78
  through bioreflex action 71
Lateral reflex equilibrium 114
Left and right extremities 52
Light 2
  dominance factor 25
  intensities 93
  interaction 37
  magnet, rotation 88
Light-blink 97
Light-Magnet-Rotation (LMR) apparatus 4, 5, 28
  LMR beam simulator 88
  LMR polar blink 100
  LMR rotor rotation 99
  LMR test apparatus 88
Locking reflex 9

M

Magnetic field of LMR's bar magnet 91
March equinox 4
Migration beam 4, 21, 22, 66, 83, 84, 99, 106
Monopodial stance 112
Motor neuro reflexes 16

N

Negative reaction 15
Neural 2
Neuro reflex deflection 16
No beam detection 17
No directional beam detection 17
Nocturnal and diurnal 33
North of magnet's neutral line 27
Northern hemisphere 22

O

Ocular antennae 55
Off-course 9
  orientation 16
  reaction 17, 30
On-course 9
  orientation 16
  reaction 14, 17, 29

P

Pivot factor 101
Pivot-to-ground factor 103, 112

Polar blink 93, 94, 100
Polar light dominance factor 25
Polarized light 92
Polaroid lenses 18
Positive reaction 15

R

Reflex deflection influence 70
Reflex guidance factor 16
Reflex intensity (BRI)
differential 65
Rotation 29
action 28
factor 2, 4, 93, 101
Rotor rotation factors 100

S

Seasonal postsolstice 84
September equinox 3
Shoreline beam 4, 21, 42, 49, 67, 76, 83
Single pole blink 99
Smoke alarm 106
Solstices 4, 94
South of equator line 27
Stress 58
Stress condition 58

U

Unblinked 97

V

Volts, AC 105

W

Waist height 8
Water beam 21, 42, 49

Z

Zeeman-Faraday effect 25, 91